Analog Circuits and Signal Processing

Series Editors:

Mohammed Ismail, Dublin, USA
Mohamad Sawan, Montreal, Canada

The Analog Circuits and Signal Processing book series, formerly known as the Kluwer International Series in Engineering and Computer Science, is a high level academic and professional series publishing research on the design and applications of analog integrated circuits and signal processing circuits and systems. Typically per year we publish between 5–15 research monographs, professional books, handbooks, edited volumes and textbooks with worldwide distribution to engineers, researchers, educators, and libraries.

The book series promotes and expedites the dissemination of new research results and tutorial views in the analog field. There is an exciting and large volume of research activity in the field worldwide. Researchers are striving to bridge the gap between classical analog work and recent advances in very large scale integration (VLSI) technologies with improved analog capabilities. Analog VLSI has been recognized as a major technology for future information processing. Analog work is showing signs of dramatic changes with emphasis on interdisciplinary research efforts combining device/circuit/technology issues. Consequently, new design concepts, strategies and design tools are being unveiled.

Topics of interest include:

Analog Interface Circuits and Systems;

Data converters;

Active-RC, switched-capacitor and continuous-time integrated filters;

Mixed analog/digital VLSI;

Simulation and modeling, mixed-mode simulation;

Analog nonlinear and computational circuits and signal processing;

Analog Artificial Neural Networks/Artificial Intelligence;

Current-mode Signal Processing;

Computer-Aided Design (CAD) tools;

Analog Design in emerging technologies (Scalable CMOS, BiCMOS, GaAs, heterojunction and floating gate technologies, etc.);

Analog Design for Test;

Integrated sensors and actuators;

Analog Design Automation/Knowledge-based Systems;

Analog VLSI cell libraries;

Analog product development;

RF Front ends, Wireless communications and Microwave Circuits;

Analog behavioral modeling, Analog HDL.

More information about this series at http://www.springer.com/series/7381

Mustafijur Rahman • Ramesh Harjani

Design of Low Power Integrated Radios for Emerging Standards

 Springer

Mustafijur Rahman
Intel Labs
Hillsboro, OR, USA

Ramesh Harjani
Department of Electrical & Computer
Engineering
University of Minnesota
Minneapolis, MN, USA

ISSN 1872-082X   ISSN 2197-1854   (electronic)
Analog Circuits and Signal Processing
ISBN 978-3-030-21335-0   ISBN 978-3-030-21333-6   (eBook)
https://doi.org/10.1007/978-3-030-21333-6

This Springer imprint is published by the registered company Springer Nature Switzerland AG.
The registered company address is: Gewerbestrasse 11, 6330 Cham, Switzerland

To my dear parents. . . .

Preface

In this book, circuit techniques pertinent to low power CMOS integrated radio design compatible with IEEE 802.15.6 standard are presented. Low power radios are in increasing demand with the advent of an era of the "Wireless Body Area Networks" and "Internet of Things". The performance of the proposed techniques have been verified by fabricating them in two standard CMOS processes: TSMC's 65 nm and IBM's 130 nm process. These designs are compatible with all the channels defined in IEEE 802.15.6 standard in the frequency range of 2.36–2.484 GHz.

First, an IEEE 802.15.6 compliant 2360–2484 MHz multiband transmitter is presented that digitally multiplexes the appropriate phases from an 800 MHz poly-phase filter output to generate $\pi/4$ DQPSK signals at 2.4 GHz using injection locking. Modulation at one-third the RF frequency reduces the transmitter power consumption and enables channel selection using an integer N PLL running at 800 MHz. The modulation technique does not require phase calibration and resolves the problems of traditional injection lock-based modulators. The prototype transmitter implemented in IBM's 130 nm technology consumes 2.4 mW while delivering -10 dBm RF power at the TX output resulting in an energy efficiency of 2.5 nJ/bit at 1.2 Mbps raw data rate. The measured RMS EVM for $\pi/4$ DQPSK modulation is 3.21%.

Second, a 2.3–2.5 GHz low power low-noise 0.7 V mixer-first RF frontend for an IEEE 802.15.6 narrowband receiver is presented which uses frequency translated mutual noise cancellation based on passive coupling. Unlike traditional noise cancelling techniques, we perform symmetrical noise cancellation of a fully differential structure where each path cancels the noise of the other at IF. This prototype design realized in TSMC's 65 nm CMOS tackles the noise figure and power consumption problems of sub-1 V mixers. The figure of merit (FOM) is 10 dB higher, and the power consumption is 194 μW which is 0.5× lower than the state of the art. The local oscillator (LO) power used is only -14 dBm.

Third, a 0.7 V low power LNA combines a 1:3 frontend balun with dual-path noise and nonlinearity cancellation for improved noise performance at low power. In traditional techniques, only the noise of the main path is cancelled, while the noise of the auxiliary path is minimized by using high power. In the proposed design, the noise and nonlinearity of both the main and the auxiliary paths are mutually cancelled, allowing for low power operation. The 2.8 dB NF -10.7 dBm IIP3 LNA in TSMC's 65 nm GP process consumes 475 μW of power resulting in an FOM of 28.8 dB which is 8.2 dB better than the state of the art.

Finally, we present an 802.15.6 compliant 2.36–2.484 GHz multiband transceiver that uses an energy-efficient programmable digital power amplifier on the transit side and a zero power passive voltage gain frontend using a 1:3 balun on the receive side to achieve low power operation. A seventh harmonic injection locked oscillator and zero power passive polyphase filter generate the phases at 2.4 GHz required for phase modulation on the transmit side and for LO generation on the receive side. This enables channel selection using a 342.86 MHz PLL, i.e., at one-seventh of the RF frequency of 2.4 GHz to result in low power consumption. The prototype transmitter consumes 1.48 mW of power while delivering -9.47 dBm output power resulting in an energy efficiency of 1.52 nJ/bit at 971 kbps data rate. The measured RMS EVM for $\pi/4$ DQPSK modulation is 5.68%. The prototype receiver consumes 1.29 mW of power resulting in an energy efficiency of 1.32 nJ/bit while achieving a receiver noise figure of 10.2 dB and an IIP3 of -24.1 dBm. This design does not use offchip inductors.

Hillsboro, OR, USA   Mustafijur Rahman
Minneapolis, MN, USA   Ramesh Harjani

Acknowledgments

I have received immense support, inspiration, and guidance from several people to reach this stage in life, and this book shall be incomplete without expressing gratitude to them.

First and foremost, I express my most sincere gratitude to my advisor Prof. Ramesh Harjani for his guidance and motivation throughout my PhD. I am fortunate to have him as an advisor who provided me the freedom to explore on my own and at the same time guided me when I was struggling with a problem. I am grateful to him for facilitating me with a fabrication of circuits in advanced technology nodes and a well-equipped lab with test instruments. Furthermore, he has helped me acquire skills related to technical writing and making quality presentations. I am also thankful to Savita Harjani for the warm hospitality and delicious food at the get-together dinner parties at their residence which I shall miss in the future.

I would like to thank Prof. F.A. Talukdar and Dr. K.L. Baishnab for supervising my undergraduate final year project at NIT Silchar. Their support has been instrumental in publishing my undergraduate research work in analog circuit design. Furthermore, I was fortunate to earn a summer research position under Prof. Roy. P. Paily at IIT Guwahati where I was exposed to MEMS and analog circuit design using state-of-the-art CAD tools.

I would like to thank my lab-mates Martin Sturm and Mohammad Elbadry for introducing me to the steps of completing a successful tapeout in silicon. They also helped me in using the test equipments in the laboratory. I am thankful to Mohammad Elbadry for helping me in the layout of digital baseband section in the transmitter and guiding me through EM simulation steps. I can never forget Taehyoun Oh for his words of inspiration which helped in fostering strong confidence within myself throughout my PhD. I am also thankful to Anindya Saha, Saurabh Chaubey, Rakesh Kumar Palani, Hundo Shin, Xingyi Hua, and Zhiheng Wang for being great friends and lab-mates.

Furthermore, I would like to thank the people in the ECE Department whose support enable graduate students to conduct research smoothly. I would like to thank Carlos Soria and Chimai Nguyen for their support in maintaining the servers, softwares, and CAD tools. I would also like to thank Linda Bullis, Dan Dobrick,

Jim Aufderhar, and Linda Jagerson for their help in purchasing components and in administrative issues.

Outside academics, I would like to thank Sri Sunil Barman in my hometown, Abhayapuri, in Assam, India, who was a retired laboratory demonstrator in a local Science College. He wrote a couple of books for designing portable radios and hobby projects using discrete components and used to present electronic projects in local science exhibitions. Being written by someone in the same town, I got excited and followed those books when I was in class VIII in school. I started designing interesting electronic projects like radio receiver and transmitter, automatic water tap using light-dependent resistor (LDR), power backup inverters, etc. Being fascinated at an early age, I decided to pursue Electronics and Communication Engineering during my undergraduate studies at NIT Silchar.

Finally, I am ever grateful to my parents for their trust and moral support. They have given me absolute freedom to pursue what I liked the most. Their blessings and good wishes have been invaluable in accomplishing my achievements.

I sincerely thank you all!

Contents

1 Introduction 1
   1.1 Organization 4
   References 4

2 Transmitter 5
   2.1 Introduction 5
   2.2 System Overview 6
   2.3 Transmitter Specifications 7
      2.3.1 System Level Specifications 7
      2.3.2 Circuit Level Specifications 8
   2.4 Circuits 10
   2.5 Analysis 15
   2.6 Process Variation and Device Mismatches 16
   2.7 Measurements 17
   2.8 Conclusion 20
   References 21

3 Receiver 23
   3.1 Introduction 23
   3.2 System Overview 24
   3.3 Receiver Specifications 25
      3.3.1 System Level Specifications 26
      3.3.2 Circuit Level Specifications 26
   3.4 Circuit Design 27
      3.4.1 Signal Path 28
      3.4.2 Noise Path 29
      3.4.3 Noise Cancellation Ratio 29
      3.4.4 Noise Analysis 33
   3.5 Noise Cancellation Simulations 34
   3.6 Impact of Process Variation 34
   3.7 Measurement Results 35
   3.8 Conclusion 39
   References 39

4   Dual-Path Noise Cancelling LNA 41
　　4.1   Introduction 41
　　4.2   Circuit Design 44
　　　　4.2.1   CS Noise Cancellation.. 44
　　　　4.2.2   CG Noise Cancellation 46
　　　　4.2.3   Zin 46
　　4.3   Signal, Noise, and Nonlinearity Analysis............................. 47
　　　　4.3.1   Signal Analysis 47
　　　　4.3.2   Noise Analysis 48
　　　　4.3.3   Nonlinearity Analysis 49
　　4.4   Impact of Process Variation 50
　　4.5   Measurement Results 50
　　4.6   Conclusions 55
　　References 55

5   Transceiver 57
　　5.1   Introduction 57
　　5.2   System Overview 58
　　5.3   Circuit Diagram 58
　　5.4   Measurement Results 60
　　5.5   Conclusion 64
　　References 65

6   Conclusions 67

Index 69

List of Figures

Fig. 1.1   Wireless body area network 2

Fig. 1.2   Untethered patient monitoring using wireless body area network... 2

Fig. 1.3   The internet of things scenario (Source: https://www.cis. com.au/blog/internet-of-things/) 3

Fig. 1.4   Prediction of 50 billion connected things/devices using internet by 2020 (performed by Cisco) 3

Fig. 2.1   Block diagram of the proposed low power transmitter 7

Fig. 2.2   Circuit diagram of eight-phase polyphase filter and MUX 10

Fig. 2.3   Phase transitions at 800 MHz and 2.4 GHz for $\pi/4$ DQPSK modulation 11

Fig. 2.4   Circuit model of ILO and plot of ϕ_{ss} vs. $\omega_0 - \omega_{inj}$ 11

Fig. 2.5   Circuit diagram of pulse slimmer, ILO, and class AB PA 12

Fig. 2.6   Amplitude of the 3rd harmonic and the ratio of 3rd harmonic and fundamental 13

Fig. 2.7   Simulated output of the pulse slimmer showing Vinj1 and Vinj2 ... 14

Fig. 2.8   Traditional injection locked technique vs. proposed technique...... 14

Fig. 2.9   Phasor plot for EVM analysis 15

Fig. 2.10   Monte Carlo simulation results of pulse slimmer and polyphase filter with MUX 16

Fig. 2.11   Monte Carlo simulation results of center frequency of ILO and PA chain 17

Fig. 2.12   Die-micrograph of the TX fabricated in IBM 130 nm CMOS 17

Fig. 2.13   Test setup of the transmitter 18

Fig. 2.14   Power consumption distribution of the transmitter.................. 18

Fig. 2.15   Measured EVM of transmitter at output power level of −9.46 dBm 18

Fig. 2.16   TX output spectrum showing ACPR of −33.34 dB 19

Fig. 2.17   Wideband TX output spectrum and transmit mask.................. 19

Fig. 2.18   Power vs. frequency of existing PLLs in the literature............... 20

Fig. 3.1   Block diagrams for (a) traditional switching mixer
            and (b) traditional noise cancellation technique 24
Fig. 3.2   Block diagram for proposed design using FTMNC 25
Fig. 3.3   Circuit diagram of the FTMNC mixer with signal addition 28
Fig. 3.4   Equivalent circuit model for M1's current noise transfer
            function and the flow diagram for the noise cancellation
            mechanism 30
Fig. 3.5   Simplified layout and circuit model for the center-tapped
            symmetric inductor acting as an inductor for differential
            signal but as a transformer for single ended noise current 31
Fig. 3.6   (a) M1's single ended noise current undergoing resistive
            division through the transformer. (b) Simplified model of
            the noise current division 31
Fig. 3.7   Simulation of noise cancellation ratio (NCR) vs upconverted
            source resistance (R) 32
Fig. 3.8   Simulation of coupling coefficient (k), quality factor (Q),
            self-inductance (L), and mutual inductance (M) of the
            center-tapped symmetric differential inductor vs. frequency 32
Fig. 3.9   (a) NTF paths from channel noise source of M1, (b) STF
            paths from PORT1, (c) NTF curves from channel noise
            source of M1, and (d) STF curves from PORT1 35
Fig. 3.10   Monte Carlo simulation results of noise figure including
            process variation and device mismatch 35
Fig. 3.11   Monte Carlo simulation results of gain including process
            variation and device mismatch 36
Fig. 3.12   (a) Process corner simulation of gain and NF vs temperature.
            (b) Measured NF and gain over 5 samples 36
Fig. 3.13   Die-micrograph of the receiver frontend 37
Fig. 3.14   Test setup of the receiver frontend 37
Fig. 3.15   Measured and simulated (a) conversion gain, S11,
            and (b) NF vs. RF frequency 37
Fig. 3.16   Measured IIP3 and two tone test output spectrum 38
Fig. 3.17   Chart comparing FOM, LO power, and noise figure 39

Fig. 4.1   Cisco's prediction of connected devices per person by 2020 42
Fig. 4.2   Traditional noise cancelling (NC) LNA and their shortcomings 43
Fig. 4.3   Coupling of traditional CS and CG noise cancelling (NC)
            LNA stages to form a coupled CS-CG NC LNA 45
Fig. 4.4   Noise cancellation mechanism in the proposed LNA 45
Fig. 4.5   Circuit model for the balun 47
Fig. 4.6   Insertion loss and coupling between secondaries of the balun 47
Fig. 4.7   Simplified model for noise cancellation of M2 48
Fig. 4.8   Modelling nonlinearity for both the CS and the CG paths 49
Fig. 4.9   Monte Carlo simulations for noise figure: process
            variation impact 50

Fig. 4.10   Monte Carlo simulation results for gain: process
            variation impact 51
Fig. 4.11   Process corner simulation results for gain and noise figure
            vs temperature 51
Fig. 4.12   Die-micrograph of the LNA 51
Fig. 4.13   Test setup for LNA measurement 52
Fig. 4.14   Measured and simulated NFs with full and partial NC 53
Fig. 4.15   Measured gain and S11 of the LNA 53
Fig. 4.16   Two tone output spectrum with full and partial cancellation 54
Fig. 4.17   FOM, power, and noise figure comparison 54

Fig. 5.1   System block diagram of the transceiver 58
Fig. 5.2   Overall circuit details for the proposed transmitter 59
Fig. 5.3   Conceptual circuit block diagram for the PA 60
Fig. 5.4   Receiver frontend circuit details 61
Fig. 5.5   Die-micrograph for the proposed transceiver 61
Fig. 5.6   Measured EVM for the transmitter 62
Fig. 5.7   Measured ACPR for the transmitter 62
Fig. 5.8   Measured gain and noise figure for the receiver 63
Fig. 5.9   Power vs frequency of existing PLLs in the literature 63

List of Tables

Table 2.1 Performance comparison of the transmitter 21

Table 3.1 Performance comparison of the receiver frontend 38

Table 4.1 Performance comparison of the LNA 54

Table 5.1 Performance comparison of the transceiver 64

Chapter 1
Introduction

A wireless body area network (WBAN) is a network of medical devices on, in, or around the human body employing wireless connectivity. WBAN promises to revolutionize health care in the near future and has gained momentum. A typical WBAN scenario is shown in Fig. 1.1 which includes medical devices such as sensors for monitoring vital data such as blood pressure, heart rate, and electrocardiogram (ECG), and actuators such as insulin pumps and cardiac pacemakers. By integrating these devices with a local control unit, e.g., a cell phone, WBAN will provide doctors with real-time data and enable remote patient monitoring. This will lead to reduced health care cost and early detection and prevention of diseases. Remote patient monitoring will significantly benefit the aging population in regions where there is a scarcity of clinics and clinicians. Furthermore, WBAN can facilitate untethered patient monitoring schemes by virtue of its wireless connectivity as shown in Fig. 1.2. Consequently, patient movement will not be limited because there will be no wires connected to those devices.

However, all these devices require radios which can transmit and receive signals in order to maintain wireless connectivity. Being powered by small batteries, these radios are power constrained and therefore they have to be extremely low power. The IEEE 802.15.6 standard provides the necessary specifications of these low power radios [1].

With the proliferation of wireless technology in the last two decades, a plethora of things have internet connectivity as shown in Fig. 1.3. To date, the world has deployed approximately five billion smart connected things. It has been predicted by Cisco (Fig. 1.4) that 50 billion things will be connected by 2020 [2]. In Fig. 1.4, these things are shown as connected devices. This has led to the so-called internet of things (IoT), which is a network of objects, animals, people, or anything provided with an IP address and the ability to transfer data without human-to-human or human-to-computer interaction. Furthermore, the latest version of IP address mechanism, i.e., IPV6 has a large address space and thus has increased the viability of the internet of things. We can assign IPV6 address to every atom on the surface

© Springer Nature Switzerland AG 2020
M. Rahman, R. Harjani, *Design of Low Power Integrated Radios for Emerging Standards*, Analog Circuits and Signal Processing,
https://doi.org/10.1007/978-3-030-21333-6_1

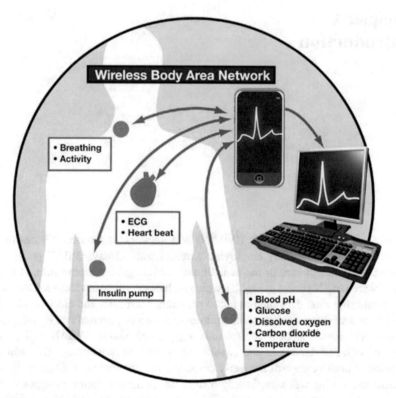

Fig. 1.1 Wireless body area network

Fig. 1.2 Untethered patient
monitoring using wireless
body area network

of earth and still have enough addresses left for another 100+ earths. However, in
order to minimize the impact of such devices on the environment and on energy
consumption, the power consumption of these IoT radios should be extremely low
when deployed for internet connectivity. In addition, many of these nodes will
not have continuous access to power and are likely to use small batteries further
increasing the need for low power radios.

Low power radios place stiff challenges on the hardware designer. Existing wireless radios are still based on homodyne or super-heterodyne conversion schemes

Fig. 1.3 The internet of things scenario (Source: https://www.cis.com.au/blog/internet-of-things/)

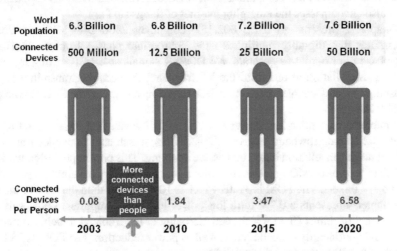

World Population	6.3 Billion	6.8 Billion	7.2 Billion	7.6 Billion
Connected Devices	500 Million	12.5 Billion	25 Billion	50 Billion
Connected Devices Per Person	0.08	1.84	3.47	6.58
	2003	2010	2015	2020

More connected devices than people

Fig. 1.4 Prediction of 50 billion connected things/devices using internet by 2020 (performed by Cisco)

invented by Edwin Armstrong in 1918 [3]. They are power hungry due to the presence of phase locked loops (PLLs), linear mixers, and digital to analog converters (DACs). The super-regenerative scheme once again invented by Armstrong in 1922 is attractive for low power implementations as it does not require a power hungry PLL and has fewer components. However, their transmission bandwidth is usually much wider than the message bandwidth making them sensitive to noise and interference. Recently introduced transmitters based on injection locking require calibration which increases production cost and time, and does not support large number of channels as needed by the 802.15.6 standard. MUX-based architectures are not suitable because generation of all the phases at RF frequency burns a huge amount of power. Finally, transmitters based on polar modulation lead to more complex circuit designs and require calibration which increases production cost and time. On the receive side, traditional architecture turns out to be power hungry due to the presence of LNA, linear mixers, and ADC. Furthermore, with scaling of technology the power supply reduction deteriorates the signal-to-noise ratio of the receive chain. Although noise cancellation techniques at the frontend may be used to restore the SNR back, traditional noise cancellation is power hungry. First, the noise of the auxiliary path is not cancelled and therefore this path burns huge power to reduce its input referred noise. Second, matching requirement calls for large transconductance of the input devices thereby increasing power consumption.

1.1   Organization

This thesis is focused on the design of low power integrated radios for wireless body area networks (WBAN) compatible with IEEE 801.15.6 standard. The application span also encompasses the emerging internet of things (IoT).

Chapter 2 presents an IEEE 802.15.6 compliant 2360–2484 MHz multiband transmitter that digitally multiplexes the appropriate phases from an 800 MHz polyphase filter output to generate $\pi/4$ DQPSK signals at 2.4 GHz using injection locking. Modulation at one-third the RF frequency reduces the transmitter power consumption and enables channel selection using an integer-N PLL running at 800 MHz.

Chapter 3 presents a low power low noise 0.7 V mixer-first RF frontend for an IEEE 802.15.6 narrowband receiver (2.3–2.5 GHz) which uses frequency translated mutual noise cancellation based on passive coupling. This prototype design realized in TSMC's 65 nm CMOS tackles the noise figure and power consumption problems of sub 1 V mixers. The FOM is 31 dB which is 10 dB higher than the state of the art.

Chapter 4 presents a 0.7 V ultra low power low noise amplifier (LNA) suitable for internet of things (IoT) which uses 1:3 balun at the frontend to achieve mutual noise and nonlinearity cancellation as well as power reduction. The FOM is 28.8 dB which is 8.2 dB better than the state of the art.

Chapter 5 presents a fully integrated transmitter and receiver system compatible with IEEE 802.15.6 standard which has an energy efficiency of 1.52 nJ/bit on transmit and 1.32 nJ/bit on receive side. The synthesizer uses 7th harmonic injection locking to drastically reduce power.

Finally, Chap. 6 provides the conclusions.

References

1. Body Area Networks, IEEE 802.15.6 (Feb 2012), http://Available:www.ieee.org
2. Cisco internet business solutions group white paper, the internet of things (April 2011) [online]
3. Who invented the superheterodyne? www.antiqueradios.com/superhet/ [online]

Chapter 2
Transmitter

2.1 Introduction

An aging population and the concern for healthcare costs have increased the desire for untethered patient monitoring of vital health signals prompting renewed efforts towards the design of low power radios for wireless body area networks (WBAN) compliant with the IEEE 802.15.6 communication standard [1]. Wireless body area networks consist of devices in, on, and around the body. These devices can be sensors for monitoring various vital physiological parameters like ECG, EEG, blood pressure, etc., or they can be actuators like pacemakers, insulin pumps, cochlear implants, etc. [2]. In addition, the WBAN transceivers may also be employed for home automation and infrastructure monitoring applications. These transceivers have short range and should consume low power as they are conceived to be powered by a coin sized battery. IEEE 802.15.6 protocol provides the necessary specifications for the transceivers employed for maintaining wireless connectivity of these devices. This paper presents an 802.15.6 multiband transmitter that is able to operate in all the 118 channels available in the new medical body area network (MBAN) band (2360–2400 MHz) and industrial scientific and medical (ISM) band (2400–2483.5 MHz). Traditional mixer-based up-conversion transmitters used in more stringent protocols are power hungry due to the presence of DACs, analog filters, linear mixers, and RF amplifier blocks [3]. Hence for low power standards, such as the IEEE 802.15.6, simpler modulation and up-conversion techniques are favored. The present state-of-the-art transmitters in this band use two point PLL-based modulation techniques and consume over 4.6 mW while transmitting [4, 5]. Transmitters based on this technique lead to more complex circuit designs that call for matching of the gain, phase, timings, and transient responses between the modulation points necessitating calibration during manufacturing and potentially during normal operation due to PVT changes. Simple RF phase multiplexing becomes power hungry at 2.4 GHz because generating all the 8 phases at such a

© Springer Nature Switzerland AG 2020
M. Rahman, R. Harjani, *Design of Low Power Integrated Radios for Emerging Standards*, Analog Circuits and Signal Processing,
https://doi.org/10.1007/978-3-030-21333-6_2

high frequency is expensive in terms of power. In this paper we propose a robust energy efficient architecture using a passive phase generation technique that utilizes sub-harmonic injection locking. This architecture avoids the limitations of recent injection locked oscillator (ILO) based PSK modulators that require calibration of the capacitor bank to achieve accurate phase shifts at the different center frequencies [6, 7]. A part of the work has been presented in [8] and we have extended this work by including design insights, derivations of circuit level specifications from the standard and more simulation and measurement results. In recent ILO techniques, the tank capacitance is varied to change the self-resonance frequency of the ILO in comparison to the injection frequency and hence a phase shift is introduced. Unfortunately, this technique needs calibration to account for process variations of the center frequency and the inherent ILO phase transition nonlinearity. Additionally, attempting to change the ILO self-resonance frequency at 2.4 GHz requires that the capacitance bank steps, ΔC, to be extremely small for generating the phases and this technique becomes impractical for supporting the large number of channels needed for the IEEE 802.15.6 narrowband PHY at 2.4 GHz. In [9] ring oscillator-based phases are multiplexed to perform direct modulation at the power amplifier achieving very good energy efficiency but the WBAN narrowband standard dictates a tighter phase noise specification which is difficult to achieve using ring oscillators. Furthermore, the prototype transmitter presented here does not use any off-chip inductors for the power amplifier unlike [5].

2.2   System Overview

The block diagram for the proposed transmitter is shown in Fig. 2.1. A MUX-based architecture is proposed where modulation occurs at 800 MHz (i.e., one-third the RF frequency). A passive polyphase filter centered at 800 MHz generates all the 8 phases necessary for $\pi/4$ DQPSK modulation at 2.4 GHz. The 8 phases generated by the polyphase filter are selected by the phase multiplexer (MUX) based on the digital baseband data. Modulation at one-third the RF frequency reduces the power consumption and also enables us to employ simpler low power circuits which are difficult to operate at the higher RF frequencies. In addition channel tuning can be achieved by a PLL running at one-third the RF frequency which further reduces power. In this prototype design the standard PLL design has not been included. However, its impact on the overall architecture and the overall power consumption has been discussed in Section VII at the end of the paper. A pulse slimmer enhances the third harmonic content at 2.4 GHz of the phase that is selected by the MUX [10]. The ILO, tuned to 2.4 GHz, locks on to this third harmonic and functions as both a high-Q bandpass filter and a frequency multiplier. Figure 2.1 also shows the constellation at 800 MHz and the constellation at 2.4 GHz that is mapped by this 3X frequency/phase multiplication.

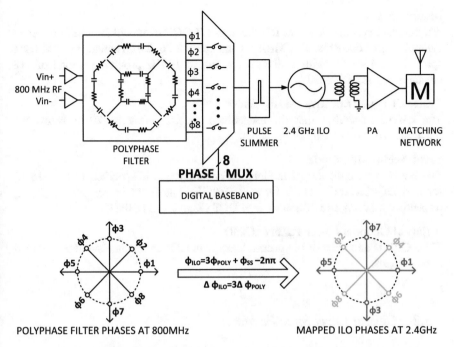

Fig. 2.1 Block diagram of the proposed low power transmitter

2.3   Transmitter Specifications

The IEEE 802.15.6 narrowband protocol encompasses the ISM band as well as the new MBAN band for medical devices in the USA. The MBAN frequency band is less crowded than ISM and hence is very suitable for medical applications as far as communications reliability is concerned [10]. The protocol supports multiple nodes and has good reliability, security, and quality of service, all of which are extremely critical for medical devices. As compared to Bluetooth LE [11], this protocol supports a higher number of channels and has higher data throughput and a longer line of sight range. In addition, this protocol supports multiple data rates which enables a trade-off between data rate and link robustness/range. The system level specifications of the transmitter as well as the circuit level specifications that were derived are discussed in the following paragraphs.

2.3.1   System Level Specifications

The system level requirements for the transmitter specified by the IEEE 802.15.6 narrowband standard [1] are briefly discussed as follows.

Modulation
The standard specifies that $\pi/2$ DBPSK and $\pi/4$ DQPSK modulation be operated at symbol rate of 600 ksps. The MBAN band spans 2.36–2.4 GHz and the ISM band spans 2.4–2.484 GHz with a channel spacing of 1 MHz providing a total of 118 channels across both bands.

Transmit Power Spectral Density Mask
The standard specifies that the transmitted spectral mask be 20 dB below the maximum spectral density of the signal.

Error Vector Magnitude
The EVM is an indicator of modulation accuracy and is specified to be -15 dB for $\pi/4$ DQPSK and -11 dB for $\pi/2$ DBPSK. The EVM in percentage (%) is, respectively, 17.7% and 28.1% for $\pi/4$ DQPSK and $\pi/2$ DBPSK.

Adjacent Channel Power Ratio (ACPR)
The ACPR is an indicator of spectral leakage in adjacent channels and is specified to be -26 dB.

2.3.2 Circuit Level Specifications

The circuit level specifications for the transmitter have been systematically derived from the transmitter requirements specified by the IEEE 802.15.6 narrowband standard described in the previous section. In particular, the derivations of the PLL phase noise and power amplifier linearity are described next.

PLL Phase Noise
The PLL phase noise is dictated by both EVM specification of the transmitter and reciprocal mixing phenomenon in the receiver. The standard specifies an EVM of -15 dB for $\pi/4$ DQPSK which is equivalent to EVM (rms) of 0.177. The EVM (rms) and integrated rms phase error are related as follows [12]:

$$\phi_{err}^2 = \frac{x}{100} \text{EVM (rms)}^2 \qquad (2.1)$$

Assuming that the phase noise of the PLL contributes 50% of the EVM, i.e., assuming $x = 50$, the integrated phase error (ϕ_{err}^2) is found to be 0.015 rad^2 using (2.1). Assuming a $1/f^2$ phase noise profile, the necessary phase noise of an integer-N PLL at an offset of Δf with a loop bandwidth f_0 is given by the following relation [12]:

$$PN(\Delta f) = 10 \log(\phi_{rms}^2) - 10 \log(4 f_0) - 20 \log(\Delta f / f_0) \qquad (2.2)$$

For an integer-N PLL with a reference frequency of 1 MHz, the loop bandwidth should be at least 10% of reference frequency, i.e., 100 KHz. Using $\phi_{rms}^2 =$

0.015, $\Delta f = 1$ MHz, $f_0 = 100$ KHz in (2.2) results in phase noise specification of -94.26 dBc/Hz at 1 MHz offset. Therefore, the phase noise specification for the 800 MHz PLL should be $20 \log(3)$ dB lower due to 3rd harmonic lock, i.e., -103.8 dBc/Hz. The phase noise of the synthesizer that satisfies the reciprocal mixing specification can be approximately estimated by the following relation [12]:

$$PN(\Delta f) = -ACR - SNR - 3 - 10 \log(\Delta f) \tag{2.3}$$

where Δf is the channel bandwidth, PN is the phase noise at an offset of Δf, and ACR (adjacent channel rejection) is the ratio of interfering signals power in the adjacent channel to the desired signal power. According to the 802.15.6 standard specification [1], the desired signals strength should be set 3 dB above the rate dependent sensitivity and the power of the interfering signal should be raised until a 10% packet error rate has been achieved for physical layer service data unit (PSDU) length of 255 octets. Using the 10% packet error rate specification at a data rate of 971.4 kbps, simulation results show that a minimum SNR of 11.2 dB is required. The specified ACR at this data rate is 9 dB. Using ACR $= 9$ dB, SNR $= 11.2$ dB, and $\Delta f = 1$ MHz for data rate of 971.4 kbps, (2.3) results in a phase noise specification of -83.2 dBc/Hz at a 1 MHz offset. Therefore, the phase noise specification of the 800 MHz PLL is $20 \log(3)$ dB lower due to 3rd harmonic lock, i.e., -92.7 dBc/Hz.

After considering all the acceptable data rates, the phase noise specification resulting from data rate of 971.4 kbps is the most stringent from the receiver side's perspective. Therefore, the 800 MHz PLL phase noise specification of -103.8 dBc/Hz from the transmitter side turns out to be more stringent and will be used as our PLL phase noise specification.

PA Nonlinearity Specification
The most strict modulation scheme in the narrowband 802.15.6 standard is $\pi/4$ DQPSK. This scheme has a maximum phase change of $135°$ and has a non-constant envelope. The peak to average ratio has been simulated using a Rohde & Schwarz WinIQSim2 software [13] and is found to be 3.2 dB. When the modulated waveform has a non-constant envelope, power amplifier nonlinearity leads to spectral regrowth which deteriorates adjacent channel power ratio (ACPR). The effect of power amplifier nonlinearity on ACPR can be quantitatively represented by the following relationship [14, 15]:

$$ACPR = IM3 + 9 - 0.85(PAR - 3) \tag{2.4}$$

where PAR is the peak to average power ratio and IM3 is the two tone third order intermodulation ratio. Using ACPR $= -26$ dB, PAR $= 3.2$ dB in (2.4) results in IM3 $= -34.83$ dBc. The required OIP3 of the power amplifier can be evaluated using the following relation:

$$IM3 = 2(Pout - OIP3) \tag{2.5}$$

Using Pout $= -10$ dBm, IM3 $= -34.83$ dBc in (2.5) results in required OIP3 requirement of approximately 7 dBm.

2.4 Circuits

The circuit details for the passive polyphase filter along with the MUX are shown in Fig. 2.2. The inner ring of the filter is a conventional polyphase filter which generates quadrature phases from a differential input signal [16]. The outer ring generates all the eight phases needed for $\pi/4$ DQPSK modulation from the quadrature waveforms [17]. The output of the polyphase filter is a low swing analog signal at 800 MHz and hence a switch-based MUX was employed to save power. When any of the switches is turned off it is terminated with a dummy buffer so as to ensure symmetric loading and prevent any phase imbalance in the polyphase filter output. The ILO phase (ϕ_{ILO}) and the phase of the polyphase filter output (ϕ_{POLY}) is related as $\phi_{ILO} = 3\phi_{POLY} + \phi_{SS} - 2n\pi$ where ϕ_{SS} is the steady-state phase difference between the injected current and the oscillator current as shown in Fig. 2.1. Therefore, phase shifts are related as $\Delta\phi_{ILO} = 3\Delta\phi_{POLY}$. As an example,

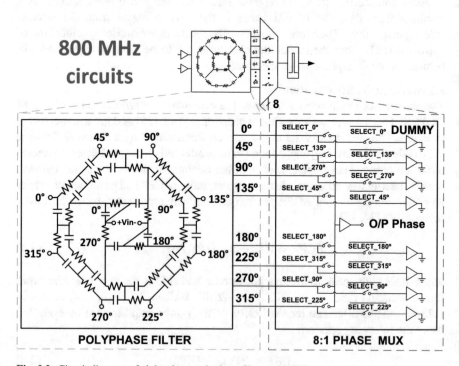

Fig. 2.2 Circuit diagram of eight-phase polyphase filter and MUX

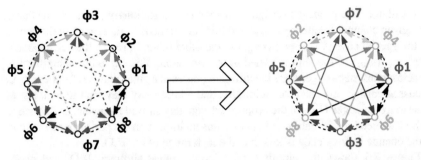

Phase transition at 800MHz   **Phase transition at 2.4GHz**

Fig. 2.3 Phase transitions at 800 MHz and 2.4 GHz for $\pi/4$ DQPSK modulation

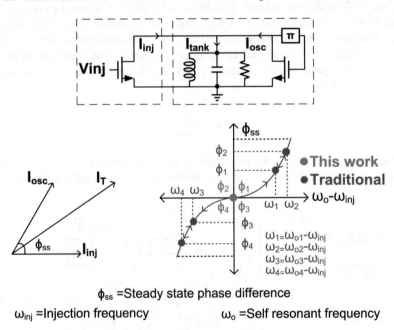

ϕ_{ss} =Steady state phase difference

ω_{inj} =Injection frequency ω_o =Self resonant frequency

Fig. 2.4 Circuit model of ILO and plot of ϕ_{ss} vs. $\omega_0 - \omega_{inj}$

a phase change of 45° causes a phase change of $3 \times 45° = 135°$ in the output. Note, this is a 1:1 mapping where the 8 phases at 800 MHz map onto the 8 required phases at 2.4 GHz as was shown in Fig. 2.1. The phase transitions at 800 MHz and corresponding transitions at 2.4 GHz for $\pi/4$ DQPSK are shown in Fig. 2.3 which shows clearly that for any required transition at 2.4 GHz there exists a corresponding unique transition at 800 MHz. Additionally, this figure shows that the transient properties of $\pi/4$ DQPSK are maintained. Figure 2.4 shows a simplified model, the phasor relationships for an ILO, and a plot of ϕ_{ss} vs. $\omega_{inj} - \omega_{srf}$ of an ILO where ω_{inj} and ω_{srf} are, respectively, the injected frequency and the self-resonant frequency of the oscillator. The proposed design corresponds to a stationary point on this curve for a given channel frequency (ω_{osc}) and this can be made to lie exactly at the center for the special case of $\omega_{srf} = 3x\omega_{inj}$. On the other hand, in [6, 7], the self-resonant frequency (ω_{srf}) of the ILO is varied to traverse along this curve by using discrete capacitor bank steps to generate the different phases at a given channel frequency. As discussed earlier, this calls for calibration and additionally may lead to lock issues at the extreme edges [18]. In the proposed design any shift in the resonant frequency of the ILO need not be calibrated out as it has no impact on the performance as long as the change in frequency is less than the lock range of the ILO.

Figure 2.5 shows the circuit details for the pulse slimmer, ILO, and class-AB power amplifier (PA). The third harmonic content of the selected phase is enhanced using the pulse slimmer [19] which is then injected onto the ILO using the two NMOS differential-pair transistors (M1, M2) as shown in Fig. 2.5. The pulse slimmer consists of a duty cycle control stage followed by a differentiator stage. The duty cycle control sets the optimal duty cycle of the pulse-slimmed signal which maximizes the 3rd harmonic content of the signal. The amplitude of the nth harmonic of a square wave with duty cycle D and unity amplitude is given by the following relation [20]:

$$y_n = \frac{\sin(n\pi D)}{n\pi} \tag{2.6}$$

For a given value of n, the amplitude y_n is maximized for several values of D because y_n is a periodic function of D. Therefore, the optimum duty cycle is selected

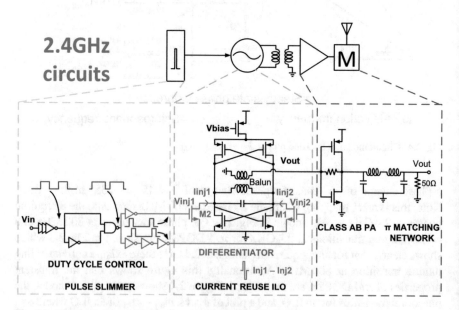

Fig. 2.5 Circuit diagram of pulse slimmer, ILO, and class AB PA

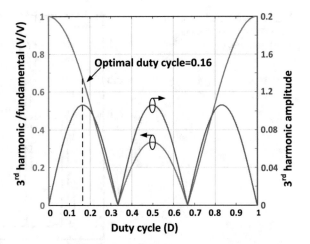

Fig. 2.6 Amplitude of the 3rd harmonic and the ratio of 3rd harmonic and fundamental

by considering additional factors. The amplitude of the fundamental also needs to be reduced in order to minimize the amount of desensitization due to the fundamental. Therefore, D may be chosen in such a way that the ratio of the 3rd harmonic and the fundamental as well as the 3rd harmonic content itself gets maximized. Figure 2.6 shows a plot of the ratio of the 3rd harmonic/fundamental as well as 3rd harmonic content itself with respect to D. The amplitude of 3rd harmonic content is periodic with period of $1/3 = 0.33$ and its maxima occurs at $D = 0.16, 0.5$, and 0.83. However, the ratio of 3rd harmonic/fundamental at $D = 0.16$ and 0.83 is greater than that at $D = 0.5$. Therefore, D= 0.16 or 0.83 may be used as the optimal duty cycle. In this design, $D = 0.16$ has been selected for hardware simplicity. The duty cycle control is implemented using a NAND gate as shown in Fig. 2.5. The 50% duty cycle square wave input is split into two paths with appropriate delay differences resulting in a $D = 0.16$ pulse after the NAND gate followed by the inverter. The pulse slimmer is followed by a discrete time differentiator which acts as a high pass filter thereby enhancing the 3rd harmonic and suppressing the lower ones. In addition, it suppresses the even order harmonics and converts the single ended signal to a pseudo differential one making it more suitable to be applied as an injection signal to the differential ILO. This is done by splitting the signal into two paths, one of which has a small additional delay implemented by inverter stages. Figure 2.7 shows the simulated pseudo differential outputs of the pulse slimmer, i.e., Vinj1 and Vinj2. Each of these outputs has a duty cycle of 0.16 and Vinj2 is delayed with respect to Vinj1 by approximately $\Delta t = 0.18$ ns using additional inverters. This delay can be adjusted in order to ensure that the lower harmonics are sufficiently suppressed.

The oscillator is a PMOS-NMOS current reuse oscillator which is inductively coupled to the power amplifier (PA) using a balun. This balun couples the differential ILO output to a single ended self-biased AC-coupled class-AB PA to further save power. The PA can operate with a VDD from 1.2 to 1.5 V and uses no off-chip inductors. The PA is followed by a wideband π matching network to

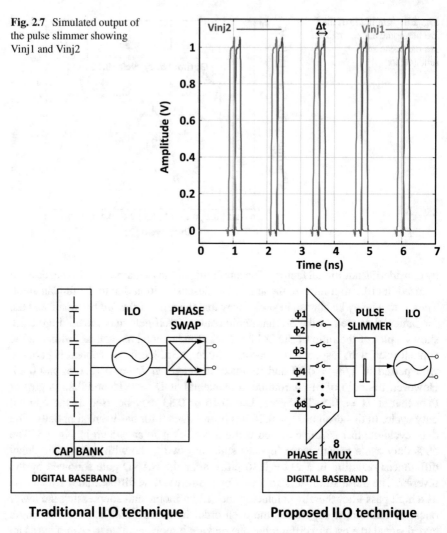

Fig. 2.7 Simulated output of the pulse slimmer showing Vinj1 and Vinj2

Fig. 2.8 Traditional injection locked technique vs. proposed technique

match a 50 ohm antenna. The circuit schematic for the PA and matching network were shown in Fig. 2.5. Figure 2.8 attempts to clarify the differences between traditional ILO techniques [6, 7] and the fixed phase selection ILO (FPS-ILO) technique introduced in this paper. In traditional ILOs the self-resonant frequency of an injection locked oscillator is varied using a capacitor bank with step capacitances, ΔC, such that there is a difference between the injection signal frequency and the ILO self-resonance frequency. This results in a phase shift in the range of $[-\pi/2$ to $\pi/2]$. A 180° phase swap circuit generates the additional phases so that the total range is $[-\pi$ to $\pi]$. The FPS-ILO technique has several benefits in comparison to traditional ILO modulation techniques. First, the phase mapping is always accurate regardless of process variations. In a polyphase filter the phases that are generated are unaffected by absolute process variations, which tend to be larger, but are only affected by the relative mismatch between components, which tend to be much smaller. Proper layout techniques as discussed in [16] have been used to minimize phase imbalance. Any amplitude imbalance is nullified by the hard switching action of the high gain inverter stage of the pulse slimmer that follows the MUX. This eliminates any calibration as in [6, 7] which has to be done for each and every channel because the capacitance step ΔC for generating each phase shift $\Delta \phi$ at different center frequencies is different. Second, selecting the appropriate phases digitally is significantly crisper and can be done at lower power at one-third the RF frequency. In addition, existing designs are not particularly suitable for higher frequencies because ΔC becomes very small to be practical and this problem worsens even further when the number of channels increases, i.e., to 118 as in the MBAN/ISM standard. Finally, this is a robust design without the process sensitive nonlinear phase-frequency mapping used in existing ILO techniques.

2.5 Analysis

As shown in Fig. 2.9, EVM is the difference in the position vector of the measured symbol and ideal symbol. The EVM (%) can be expressed in (2.7) as follows [7]:

$$
\text{EVM (\%)} = \sqrt{\left(\frac{\Delta M}{OA}\right)^2 + (\sin \Delta\phi)^2} \times 100\% \tag{2.7}
$$

where ΔM is the amplitude error, OA is the ideal phasor magnitude, and $\Delta \phi$ is the phase error. The focus of our EVM analysis is to encompass the effect of systematic constellation error due to non-idealities in the polyphase filter output. The effect of amplitude imbalance in the polyphase filter output on EVM is nullified by the hard switching action of the digital pulse slimmer and therefore $\Delta M \approx 0$.

Fig. 2.9 Phasor plot for EVM analysis

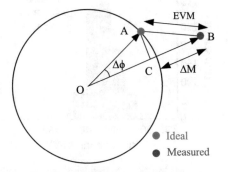

Furthermore, the effect of random noise sources is not considered to simplify the analysis. Therefore, using (2.7) EVM (%) can be expressed as $\sin(\Delta\phi) \times 100\%$. Furthermore, the modulated output of the transmitter can be written as follows:

$$V_{out} = V_o \cos(2\pi f_{ILO}t + \phi_{ILO}(t)) = V_o \cos(2\pi(3f_{INJ})t + 3\phi_{INJ}(t) + \phi_{SS})$$
(2.8)

where f_{ILO} is the frequency, $\phi_{ILO}(t)$ is the phase of the ILO, f_{INJ} is the frequency, $\phi_{INJ}(t)$ is the phase of the injection signal at the output of MUX, and ϕ_{SS} is the steady-state phase difference. Due to device mismatch and frequency deviation from the center frequency there can be small phase errors ϕ_{INJ} in the polyphase filter output and therefore in the MUX output [17]. As shown in (2.8), the ILO being locked to the third harmonic, the phase error at the ILO output is $= 3\Delta\phi_{INJ}$ and therefore the rms EVM is given by $\sin(3\Delta\phi_{INJ}) \times 100\%$.

2.6 Process Variation and Device Mismatches

Monte Carlo simulations were performed to quantify the effect of process variation and mismatch on the performance of the polyphase filter, MUX, and pulse slimmer. As shown in Fig. 2.10, the standard deviation of the phase at the output of the polyphase filter and MUX is only 1.14° with σ/μ ratio of 1.26%. This simulation result does not include the static delay of the pulse slimmer which does not vary over symbol periods and therefore gets cancelled. However, this delay does affect the third harmonic content. Therefore, the effect of process variation and mismatch on the pulse slimmer has been quantified by performing Monte Carlo simulations on the third harmonic content of the pulse slimmer output. Figure 2.10 also shows the standard deviation of the third harmonic content as 24.5 mV with σ/μ ratio of 6.5%. Finally, Monte Carlo simulations have been done on the center frequency of ILO and PA chain and standard deviation is only 11 MHz as shown in Fig. 2.11.

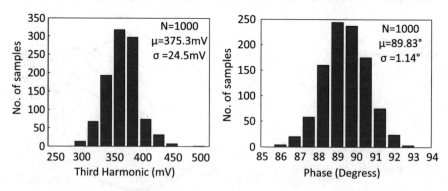

Fig. 2.10 Monte Carlo simulation results of pulse slimmer and polyphase filter with MUX

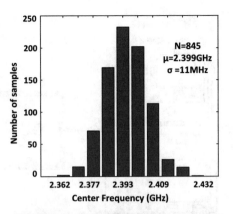

Fig. 2.11 Monte Carlo simulation results of center frequency of ILO and PA chain

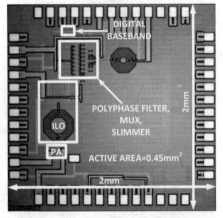

Fig. 2.12 Die-micrograph of the TX fabricated in IBM 130 nm CMOS

2.7 Measurements

The transmitter was implemented in IBM's 130 nm CMOS technology and occupies a core area of 0.45 mm². The die microphotograph is shown in Fig. 2.12 and the test setup is shown in Fig. 2.13. The 800 MHz reference signal was generated using an external signal generator. Transmission lines fabricated on the FR4 board were used for the 800 MHz reference signal input and for observing the 2.4 GHz modulated transmitter output on the spectrum analyzer. An arbitrary waveform generator was used to generate the digital baseband data. The EVM and ACPR measurements were done using a Rohde and Schwarz FSW43 spectrum analyzer [21].

The power consumption distribution is shown using a pie chart in Fig. 2.14. The power amplifier consumes 1.4 mW of DC power with a VDD of 1.5 V at −10 dBm TX output power. The ILO consumes 400 μW and the circuits at 800 MHz including drivers consume 600 μW with a VDD of 1 V. Note that circuits at 800 MHz consume only 25% of total power. The measured rms EVM for π/4 DQPSK modulation is 3.21% as shown in Fig. 2.15.

The output transmit spectrum is shown in Fig. 2.16 which shows an ACPR of −33.34 dB at a transmit power of −9.46 dBm. For this test the desired ACPR was

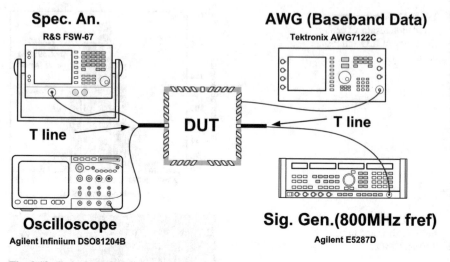

Fig. 2.13 Test setup of the transmitter

Fig. 2.14 Power consumption distribution of the transmitter

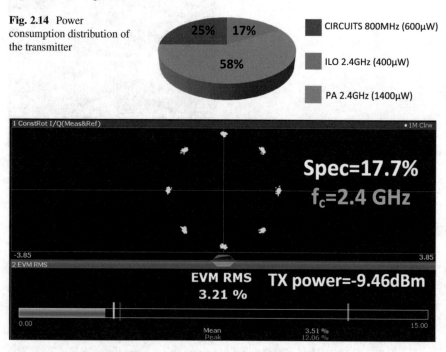

Fig. 2.15 Measured EVM of transmitter at output power level of −9.46 dBm

achieved by adjusting the VDD of the pulse slimmer and bias current of the ILO which sets the injection current (I_{inj}) and oscillator current (I_{osc}), respectively. A higher ACPR can be achieved by smoothing out the phase transitions by making

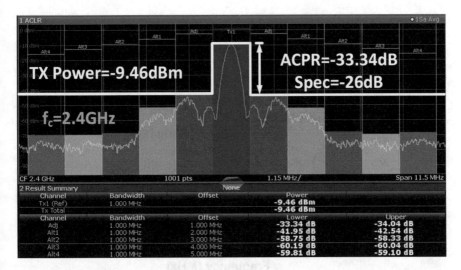

Fig. 2.16 TX output spectrum showing ACPR of −33.34 dB

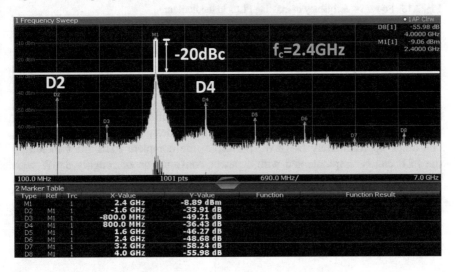

Fig. 2.17 Wideband TX output spectrum and transmit mask

I_{inj}/I_{osc} small, i.e., by limiting the filtering bandwidth of the ILO that behaves as a LPF [22].

The wideband transmit mask of the transmitter is shown in Fig. 2.17 which shows that the out of band harmonics are much below the specified transmit masks in the 802.15.6 narrowband standard. In particular, note that the 800 MHz injected frequency (D2 in Fig. 2.17) is 33.91 dB lower than the 3rd harmonic at 2.4 GHz as a result of the pulse slimming and ILO bandpass filtering. The lock range has been validated at band edges, i.e., 2360–2484 MHz.

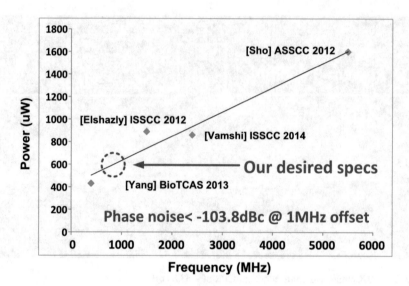

Fig. 2.18 Power vs. frequency of existing PLLs in the literature

As discussed earlier, this prototype does not include a PLL. Figure 2.18 shows the plot of the power consumption vs. the operating frequency of some recent PLLs in the literature which indicates an increasing trend of power consumption with increasing frequencies. All of these PLLs have better phase noise than the required specification of −103.8 dBc/Hz at 1 MHz offset as discussed in Section III. The plot clearly also validates the feasibility of designing a 800 MHz PLL that consumes approximately 600 μW. Therefore, the complete transmitter including the PLL can be implemented with a power consumption of roughly 3 mW with 2.5 nJ/bit efficiency which is lower than prior work [4, 5]. The ILO phase noise is −108 dBc/Hz at 1 MHz offset which has sufficient margin to meet the standard specifications. A summary of the measured results and comparison with prior work is shown in Table 2.1.

2.8 Conclusion

Modulation at 800 MHz, i.e., one-third the RF frequency using the sub-harmonic FPS-ILO technique results in a simple and low power design. The design is flexible and therefore can incorporate other harmonics provided the phase mapping is appropriately thought through. This technique may be used for M-ary PSK. Power is further reduced by using a single ended class-AB PA without any off-chip inductor. Also an integer-N PLL at 800 MHz can be employed for frequency

Table 2.1 Performance comparison of the transmitter

Reference	[4]	[5]	[7]	[23]	This work
Standard	802.15.6	802.15.6	N/A	802.15.4	802.15.6
Frequency (GHz)	2.36–2.4	2.36–2.484	915	2.405–2.48	2.36–2.484
Data rate	971 kbps	971 kbps	50 Mbps	2 Mbps	971 kbps
Modulation	$\pi/4$ DQPSK	$\pi/4$ DQPSK	QPSK	HS-QPSK	$\pi/4$ DQPSK
Technology	90 nm	130 nm	180 nm	180 nm	130 nm
Output power	−10 dBm	−10 dBm	−3.3 dBm	3 dBm	−10 dBm
EVM	7.3%	7.3%	6.41%	2%	3.21%
ACPR	−32	N/A	N/A	N/A	−33.34
Power dissipation (mW)	4.6	5.9	5.88	32.4	2.4
Energy efficiency (nJ/bit)[a]	3.8	6.07	0.12	16.2	2.5

[a]Total power including PLL

synthesis resulting in improved power efficiency. The energy efficiency including the estimated power of the PLL is 2.5 nJ/bit (1.5–6.5X improvement compared to current state of the art). The robust PVT tolerant modulation technique gets rid of the calibration needed for open loop PLL-based techniques [4, 5]. The modulation technique is very precise with an EVM of 3.21% or −29.8 dB and may be suitable for protocols with tighter specifications. Furthermore, this design does not require cap bank calibration, can support 118 channels even at high GHz frequencies, and has no nonlinear phase mapping issues unlike existing injection locking techniques [6, 7].

References

1. Body Area Networks, IEEE 802.15.6 (2012). www.ieee.org
2. H.-J. Yoo, A. Burdett, Body area network: technology, solutions, and standardization, in *IEEE International Solid State Circuits Conference* (IEEE, Piscataway, 2011), p. 531
3. P. Choi, H. Park, Ilku Nam, K. Kang, Y. Ku, S. Shin, S. Park, T. Kim, H. Choi, S. Kim, S.M. Park, M. Kim, S. Park, K. Lee, An experimental coin-sized radio for extremely low power WPAN (IEEE 802.15.4) application at 2.4 GHz, in *IEEE International Solid State Circuits Conference* (IEEE, Piscataway, 2003), pp. 92–93
4. Y.H. Liu, X. Huang, M. Vidojkovic, A. Ba, P. Harpe, G. Dolmans, H.D. Groot, A 1.9 nJ/b 2.4 GHz multi-standard (Bluetooth low energy/Zigbee/IEEE 802.15.6) transceiver for personal/body-area networks, in *IEEE International Solid State Circuits Conference* (IEEE, Piscataway, 2013), pp. 446–447
5. A. Wong, M. Dawkins, G. Devita, N. Kasparidis, A. Katsiamis, O. King, F. Lauria, J. Schiff, Alison Burdett. A 1 V 5 mA multimode IEEE 802.15.6/Bluetooth low-energy WBAN transceiver for biotelemetry applications, in *IEEE International Solid State Circuits Conference* (IEEE, Piscataway, 2012), pp. 300–301
6. S.-J. Cheng, Y. Gao, W.-D. Toh, Y. Zheng, M. Je, C.-H. Heng, A 110 pJ/b multichannel FSK/GMSK/QPSK/p/4-DQPSK transmitter with phase-interpolated dual-injection DLL-based synthesizer employing hybrid FIR, in *IEEE International Solid State Circuits Conference* (IEEE, Piscataway, 2013), pp. 450–451

7. S. Diao, Y. Zheng, Y. Gao, S.J. Cheng, X. Yuan, M. Je, C.H. Heng, A 50-Mb/s CMOS QPSK/O-QPSK transmitter employing injection locking for direct modulation. IEEE Trans. Microwave Theory Tech. **60**(1), 120–130 (2012)
8. M. Rahman, M. Elbadry, and R. Harjani, A 2.5 nJ/bit multiband (MBAN & ISM) transmitter for IEEE 802.15.6 based on a hybrid polyphase-MUX/ILO based modulator, in *IEEE International Solid State Circuits Conference* (IEEE, Piscataway, 2014), pp. 17–20
9. L. Xiayun, M. M. Izad, Y. Libin, C.-H. Heng, A 13 pJ/bit 900 MHz QPSK/16-QAM band shaped transmitter based on injection locking and digital PA for biomedical applications. IEEE J. Solid State Circuits **49**(11), 2408–2421 (2014)
10. A. Wong, M. Dawkins, G. Devita, N. Kasparidis, A. Katsiamis, O. King, F. Lauria, J. Schiff, A. J. Burdett, 1 V 5 mA multimode IEEE 802.15.6/Bluetooth low-energy WBAN transceiver for biotelemetry applications. IEEE J. Solid State Circuits **48**(1), 186–198 (2013)
11. Bluetooth SIG, specification of the bluetooth system v4.0 (2010). www.Bluetooth.org. Accessed 30 June 2010
12. B. Razavi, *RF Microelectronics* (Prentice Hall, Upper Saddle River, 2012)
13. RS WinIQSIM Software Manual, Rohde & Schwarz, www.rohde-schwarz.com
14. Y.-H. Liu, X. Huang, M. Vidojkovic, G. Dolmans, H. de Groot, An energy-efficient polar transmitter for IEEE 802.15.6 body area networks: system requirements and circuit designs. IEEE Commun. Mag. **50**(10), 118–127 (2012)
15. N.B.D. Carvalho, J.C. Pedro, Compact formulas to relate ACPR and NPR to two-tone IMR and IP3. Microw. J. **42**(12), 70–84 (1999)
16. F. Behbahani, Y. Kishigami, J. Leete, A.A. Abidi, CMOS mixers and polyphase filters for large image rejection. IEEE J. Solid State Circuits **36**(6), 873–887 (2001)
17. M. Shimozawa, K. Nakajima, H. Ueda, T. Tadokoro, N. Suematsu, An even harmonic image rejection mixer using an eight-phase polyphase filter, in *IEEE MTT-S International Microwave Symposium Digest Papers* (IEEE, Piscataway, 2008), pp. 1485–1488
18. B. Razavi, A study of injection locking and pulling in oscillators. IEEE J. Solid State Circuits **39**(9), 1415–1424 (2004)
19. M. Elbadry, B. Sadhu, J. Qiu, R. Harjani, Dual-channel injection locked quadrature LO generation for a 4-GHz instantaneous bandwidth receiver at 21-GHz center frequency. IEEE Trans. Microw. Theory Tech. **61**(3), 1186–1199 (2013)
20. B.P. Lathi, Z. Ding, *Modern Digital and Analog Communication Systems* (Oxford University Press, Oxford, 2009)
21. R&S FSW MSRA Mode User Manual, Rohde & Schwarz. www.rohde-schwarz.com
22. S. Kalia, M. Elbadry, B. Sadhu, S. Patnaik, J. Qiu, R. Harjani, A simple, unified phase noise model for injection-locked oscillators, in *IEEE Radio Frequency Integrated Circuits (RFIC) Symposium Digest Papers* (IEEE, Piscataway, 2011), pp. 1–4
23. G. Retz, A highly integrated low-power 2.4 GHz transceiver using a direct-conversion diversity receiver in 0.18 μ CMOS for IEEE802.15.4 WPAN, in *IEEE International Solid State Circuits Conference* (IEEE, Piscataway, 2009), pp. 414–415

Chapter 3
Receiver

3.1 Introduction

In this chapter, a low power low noise 0.7 V mixer-first RF frontend for an IEEE 802.15.6 narrowband receiver is presented which uses frequency translated mutual noise cancellation based on passive coupling. Unlike traditional noise cancelling techniques we perform symmetrical noise cancellation of a fully differential structure where each path cancels the noise of the other at IF. This prototype design realized in TSMC's 65 nm CMOS tackles the noise figure and power consumption problems of sub 1 V mixers. The figure of merit (FOM) is 10 dB higher and the power consumption is 194 μW which is 0.5X lower than the state of the art. The local oscillator (LO) power used is −14 dBm.

Low power IEEE 802.15.6 standard-based wireless body area network (WBAN) RF frontends are starting to proliferate as the standard matures [1–3]. A WBAN is a network of medical devices on, in, or around the human body employing wireless connectivity. WBAN promises to revolutionize health care in the near future. A typical WBAN scenario includes a network of medical sensors for monitoring vital statistics such as blood pressure, heart rate, and actuators such as insulin pumps and cardiac pacemakers. By integrating these devices with a local base unit, e.g., a cell phone, WBAN will provide doctors with real-time data and enable remote patient monitoring. This will lead to reduced health care cost and early detection and prevention of diseases. Remote patient monitoring will significantly benefit the aging population in regions where there is a scarcity of clinics and clinicians. Furthermore, the wireless connectivity can facilitate untethered patient monitoring without limiting patient movement. However, all these devices require radios which can transmit and receive signals in order to maintain wireless connectivity. As most of these devices are likely to be powered by small batteries, their radios have to be extremely power efficient. The IEEE 802.15.6 standard provides the necessary specifications of these low power radios [4].

© Springer Nature Switzerland AG 2020
M. Rahman, R. Harjani, *Design of Low Power Integrated Radios for Emerging Standards*, Analog Circuits and Signal Processing,
https://doi.org/10.1007/978-3-030-21333-6_3

As a significant percentage of these sensors and actuators have large digital content the overall power is minimized via technology scaling and employing sub 1 V RF circuits. In this paper a sub 1 V low power low noise RF downconverter for an IEEE 802.15.6 narrowband receiver that operates in all the three bands viz. industrial-scientific-medical (ISM), US medical body area networks (MBAN), and European MBAN bands spanning 2.3–2.5 GHz is presented. The traditional bottleneck for sub 1 V operation has been the mixer. Existing low voltage mixers can be classified as bulk injection, switching, or square law mixers. Bulk injection mixers need a large LO and are sensitive to process variations [5]. Switching mixers require an even larger LO close to 0 dBm making them power hungry [6–8]. Recently proposed nonlinearity-based [9] and transconductance-based mixers [10] use smaller LOs. However, these mixers have relatively high noise figures (11.2 dB [9], 19 dB [10]) and high power consumption (380 μW [9], 1 mW [10]).

3.2 System Overview

A block diagram for a traditional switching mixer is shown in Fig. 3.1a. This mixer requires a large LO to achieve a low noise figure and good conversion gain not making it suitable for low VDD operation. However, active mixers have poor noise figure (NF) due to flicker noise and thermal noise folding [11, 12] and therefore, a low noise amplifier (LNA) preceding the mixer is traditionally employed at the cost of increased power. We propose an active transconductance mixer that uses low LO power to achieve low VDD operation as shown in Fig. 3.2.

The output SNR of a system can be expressed as $SNR_{out} = \epsilon VDD^2/N_{out}$, where $0 < \epsilon < 1$ and N_{out} is the output noise power. Low VDD operation decreases SNR_{out} and hence increases the effective noise figure. Therefore, noise cancellation techniques become critical at lower VDDs.

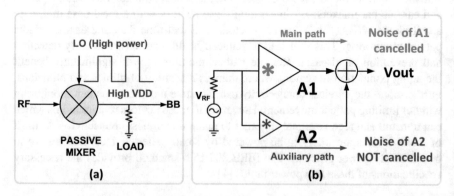

Fig. 3.1 Block diagrams for (a) traditional switching mixer and (b) traditional noise cancellation technique

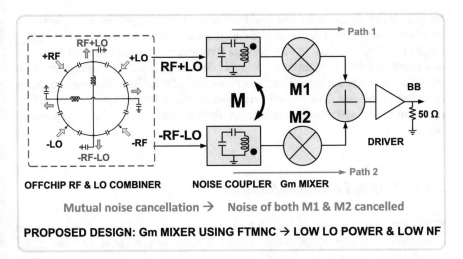

Fig. 3.2 Block diagram for proposed design using FTMNC

As shown in Fig. 3.1b, traditional noise cancelling techniques [13, 14] use an auxiliary path to cancel the noise of the main signal path but the noise of the auxiliary path still remains uncancelled. In this paper, a frequency translated mutual noise cancelling (FTMNC) mixer with subthreshold MOS operation is proposed. Unlike traditional techniques, we perform symmetrical noise cancellation of a fully differential structure where each path cancels the noise of the other side after downconversion to IF. The RF and LO is combined and applied to the mixer in differential form. A passive noise coupling mechanism couples the noise *current* of one side of the differential topology to the other side thereby making it common mode but retains the signal *voltage* in differential mode. Each end has a transconductance mixer which downconverts the RF to baseband by generating in-phase IF signals which are added at the baseband. Noise from each half, on the other hand, is common mode and gets downconverted out of phase (by +LO and −LO) and hence gets cancelled due to the addition after frequency translation [14]. In this prototype the RF and LO combination is done off-chip for testing ease. In future, a passive zero power cyclic combiner used in [9] may be used for an on-chip implementation.

3.3   Receiver Specifications

The IEEE 802.15.6 narrowband protocol [15] includes the ISM band as well as the new US MBAN and EU MBAN band for medical devices. The MBAN frequency band is less crowded than ISM and hence is particularly suitable for medical applications where reliability is critical. The protocol has been designed for good reliability, security and quality of service, and support for multiple nodes, all of which are critical for medical devices. As compared to Bluetooth low energy (BLE) [16], this protocol has higher data throughput, a longer line of sight range, and supports a higher number of channels. We propose a low IF architecture for the receiver because a zero IF architecture suffers from flicker noise and requires dc offset correction. The system level specifications for the receiver as well as the circuit level specifications that were derived are discussed in the following paragraphs.

3.3.1 System Level Specifications

IEEE 802.15.6 narrowband standard [4] was recently ratified and therefore we enlist the system level specifications for the receiver. In particular, we focus on the specifications that affect receiver RF circuit performance. In the next sub-section we translate these system level specifications to circuit and block level specification for the receiver.

- Modulation and frequency range: The standard specifies $\pi/2$ differential quadrature phase-shift keying (DQPSK) and $\pi/4$ DQPSK modulation operated at symbol rate of 600 ksps as the modulation schemes. The US MBAN band spans from 2.36 to 2.4 GHz and the US FCC Part 15 unlicensed 2.4 GHz band spans from 2.4 to 2.4835 GHz with a 1 MHz channel spacing providing a total of 118 channels across both bands. The standard also includes the EU MBAN which spans from 2.484 to 2.5 GHz.
- Receiver sensitivity: The strictest sensitivity specification is -92 dBm at a data rate of 121.4 kbps and the lowest is -83 dBm at data rate of 971.4 kbps.
- Adjacent channel rejection (ACR): The highest ACR specification is 17 dB for 121.4 kbps data rate and lowest specification is 9 dB for a 971.4 kbps data rate.

3.3.2 Circuit Level Specifications

The circuit level specifications for the receiver have been systematically derived from the system requirements specified by the IEEE 802.15.6 narrowband standard described in the previous section. In particular, a detailed derivation for the receiver noise figure and linearity are described next.

Noise Figure
The required noise figure of the receiver can be evaluated using the following equation [17]:

$$\text{Sensitivity} = -174 + \text{NF} + 10\log(\text{BW}) + \text{SNR} \tag{3.1}$$

where NF is the noise figure, BW is the channel bandwidth, i.e., 1 MHz, and SNR is the required signal-to-noise ratio. At data rate of 971.4 kbps the specified sensitivity is −92 dBm and the SNR needed is 11.2 dB. Using these values in (3.1) results in maximum NF of 19.2 dB. After considering all the acceptable data rates, this NF specification resulting from data rate of 971.4 kbps is the most stringent. We further provide margin for implementation loss in the RF frontend and digital baseband and therefore target a noise figure of about 10 dB.

Linearity

Adjacent channel rejection is the ratio of the interfering signals power in the adjacent channel to the desired signal power. According to the 802.15.6 standard specification [4], the desired signals strength should be set 3 dB above the rate dependent sensitivity and the power of the interfering signal should be raised until a 10% packet error rate has been achieved for physical layer service data unit (PSDU) length of 255 octets. At data rate of 971.4 kbps the specified ACR is 9 dB, sensitivity is −83 dBm, and the SNR required is 11.2 dB. We provide a margin of 10 dB to account for other non-idealities and therefore target an SNR of 21.2 dB. This sets a limit for third order intermodulation (IM3) corruption to be −21.2 dB. We represent the desired, adjacent, and alternate channel by $A_0 \cos \omega_0 t$, $A_1 \cos \omega_1 t$, and $A_2 \cos \omega_2 t$, respectively. Therefore, IM3 corruption [17] may be expressed as follows in (3.2):

$$20 \log \frac{3\alpha_3 A_1^2 A_2}{4\alpha_1 A_0} = -21.2 \tag{3.2}$$

$$\text{IIP3 (dBm)} = 20 \log \sqrt{\frac{4\alpha_1}{3\alpha_3}} \tag{3.3}$$

Using an ACR = 9 dB and a sensitivity = −83 dBm in (3.2), we find the value for α_3/α_1. Using this value in (3.3) we find that the required input third order intercept point (IIP3) is −55.9 dBm. After considering all the acceptable data rates, this IIP3 resulting at a data rate of 971.4 kbps is the most stringent. Since the IIP3 value dictated by adjacent channel specifications is quite relaxed we choose our target IIP3 = −19 dBm to meet the more stringent out of band blocker specifications [2].

3.4   Circuit Design

The circuit level implementation is shown in Fig. 3.3. The transconductance mixer is implemented by using an NMOS transistor in common gate configuration with a resistive load. The transistor is self-biased via gate-to-drain resistive feedback and maintained in subthreshold for a VDD = 0.7 V. The gate is ac grounded with a large capacitor. The input RF + LO and −RF − LO is applied in true differential form using an on-chip matching network implemented with a center-tapped symmetric differential inductor which also acts as a noise coupler as

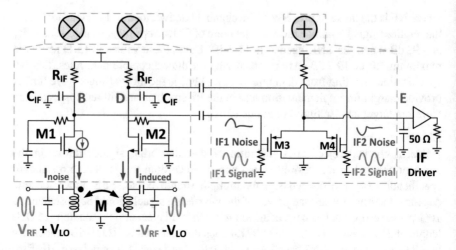

Fig. 3.3 Circuit diagram of the FTMNC mixer with signal addition

discussed later. Furthermore, the inductor tunes out the C_{gs} of the transistors and improves the transistor f_T which otherwise degrades in subthreshold. The circuit behavior with respect to signal and noise is discussed separately next.

3.4.1 Signal Path

Driven by the LO, the gm of transistors M1 and M2 is linear time variant and periodic with LO frequency and may be expressed as a Fourier series as follows:

$$gm1 = g0 + g1 \cos \omega_{LO} t + \cdots \tag{3.4}$$

$$gm2 = g0 - g1 \cos \omega_{LO} t + \cdots \tag{3.5}$$

$$id1 = +g1 \cos \omega_{LO} t \cdot (+V_{RF}) = +g1 V_{RF} \cos \omega_{LO} t \tag{3.6}$$

$$id2 = -g1 \cos \omega_{LO} t \cdot (-V_{RF}) = +g1 V_{RF} \cos \omega_{LO} t \tag{3.7}$$

As shown in (3.4) and (3.5), the fundamental harmonics of $gm1$ and $gm2$ are of opposite sign because the LO is applied in differential form. Since the RF signals are also applied in differential form, both $+V_{RF}$ and $-V_{RF}$ get downconverted to the positive quantity $+g1 V_{RF} \cos(\omega_{LO} t)$ at IF as shown in (3.6) and (3.7). As shown in Fig. 3.3 the downconverted signals are in phase and when summed together result in signal addition at the IF output. Due to differential form of input signal, the center-tapped symmetric inductor behaves as a differential inductor and forms a matching network at the input.

3.4.2 Noise Path

The noise of the transistors M1 and M2 is cyclostationary in nature, i.e., noise has a periodically time varying statistics. This occurs due to the presence of a large periodically time varying LO signal which affects noise in two ways. First, M1 and M2 have a periodically time varying operating point which modulates the noise source; and second, the noise transfer function from the noise source to the output is periodically time varying which leads to aliasing in the frequency domain [18, 19]. Consequently there is a strong correlation in the noise power spectral density at frequencies separated by integral multiples of ω_{LO}.

The equivalent circuit model for M1's noise transfer function is shown in Fig. 3.4. The wide sense stationary (WSS) noise current of M1 at A gets converted to cyclostationary noise at B due to the presence of +LO resulting in strong correlation at frequencies separated by ω_{LO}. This noise current acts as a single ended excitation to the center-tapped symmetric inductor which now behaves as a transformer. A simplified layout of the transformer and an equivalent direct model [20, 21] is shown in Fig. 3.5. I_{noise} *entering* terminal P of the primary side induces an opposite current $I_{induced}$ in the adjacent secondary trace according to Faraday's law which, however, *enters* terminal S due to the interleaved winding pattern. Consequently, noise current of M1 becomes common mode in both ends of the differential architecture. The induced current can be expressed as $I_{induced} = M_{21} \cdot I_{noise}/L_S$ where M_{21} and L_S are the mutual and self-inductances of the secondary, respectively. Furthermore, as shown in Fig. 3.4, the noise also gets bandpass filtered at RF frequency at point C due to bandpass nature of the tuned transformer [20]. The bandpass filtered noise of M1 at C gets downconverted to IF at D, as shown in Fig. 3.4. However, being downconverted by −LO, noise at D is negatively correlated with the noise at B. The noise at B and D when summed together results in noise cancellation at the IF output at E. Thus the sign reversal of the LOs applied to the left and right provides the phase reversal needed for noise cancellation. Note that there is *mutual cancellation* where the noise of M2 is cancelled in a similar fashion. This becomes possible because the transformer acting as noise coupler is a *passive reciprocal network*. Unlike active noise cancellation that uses an active but noisy auxiliary path [13, 14, 22], our passive approach consumes *zero power* and has *negligible noise* contribution. The baseband adder is implemented using PMOS source followers (M3 and M4) with a resistive load to reduce flicker noise. The gates are biased to ground with resistors. The single ended baseband output is buffered by an on-chip driver with 50 Ω matched output to drive test instruments.

3.4.3 Noise Cancellation Ratio

As shown in Fig. 3.6a the noise current of M1 acts as a single ended excitation to the center-tapped differential inductor with its center tap connected to ground.

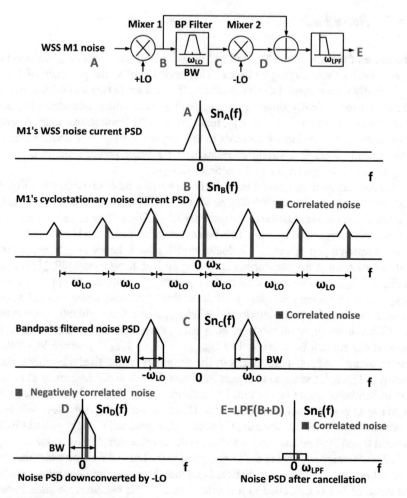

Fig. 3.4 Equivalent circuit model for M1's current noise transfer function and the flow diagram for the noise cancellation mechanism

Therefore the inductor acts as a transformer. The grounded center tap prevents the direct noise of M1 to flow to the secondary connected to M2. Instead an induced noise current flows in the secondary by virtue of mutual inductance. The impedances on the secondary side, i.e., $1/g_m$ looking into M2 and the upconverted source resistance R are now visible at the primary side due via the mutual inductive coupling. Therefore, the noise current of M1 undergoes a resistive division in between $1/g_m$ of M1, $1/g_m$ of M2, upconverted source impedance R of primary, and upconverted source impedance R of secondary as shown in the simplified circuit model in Fig. 3.6b. In this design, $R \approx 5Rs$ where Rs is the source impedance. The ratio of noise currents, i.e., I_2/I_1 or the noise cancellation ratio (NCR) may be expressed as follows:

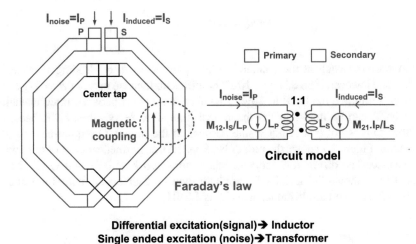

Differential excitation(signal)➜ Inductor
Single ended excitation (noise)➜Transformer

Fig. 3.5 Simplified layout and circuit model for the center-tapped symmetric inductor acting as an inductor for differential signal but as a transformer for single ended noise current

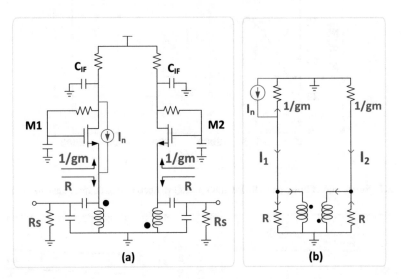

Fig. 3.6 (a) M1's single ended noise current undergoing resistive division through the transformer. (b) Simplified model of the noise current division

$$NCR = I_2/I_1 = \frac{R}{\frac{2}{g_m} + R} \tag{3.8}$$

A perfect match at the frontend will require $R = 1/g_m$ leading to NCR of 33.33%. However, if an S11 of -10 dB is sufficient, then we may set $R = 2/g_m$ and this will improve the NCR to 50%. Therefore, a trade-off between input matching and NCR can be made. In the theoretical limit if $R = \infty$, then NCR becomes 100%. This is not practically viable because of the infinite Q requirement for the inductor. Figure 3.7 shows the plot of NCR vs the transformed value R of the source resistance. The differential inductor has a simulated k of 0.87 and Q of 17.6 at 2.4 GHz as shown in Fig. 3.8a. Figure 3.8b shows the simulated mutual inductance (M) as 2.8 nH and a self-inductance (L) as 3.2 nH.

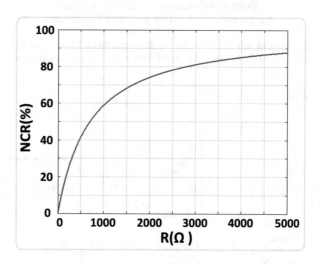

Fig. 3.7 Simulation of noise cancellation ratio (NCR) vs upconverted source resistance (R)

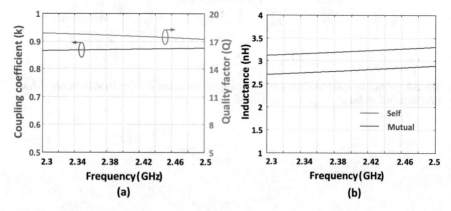

Fig. 3.8 Simulation of coupling coefficient (k), quality factor (Q), self-inductance (L), and mutual inductance (M) of the center-tapped symmetric differential inductor vs. frequency

3.4.4 Noise Analysis

The drain current of M1 and M2 biased in saturation in the subthreshold region [23] is given by the following expression:

$$I_d = \frac{W}{L} I_{do} \cdot e^{\frac{V_{GS}}{nU_T}} \tag{3.9}$$

where n, I_{do} are process dependent parameters and $U_T = KT/q$. This mixer exploits the exponential dependence of drain current on V_{GS}; and therefore, the conversion gain of the mixer can be evaluated by using the second order term in the Taylor series expansion of the exponential in (3.9). The conversion gain of a single ended stage excluding the matching network gain is expressed as follows:

$$A_{mix1} = \frac{W}{2L} \frac{I_{do}}{n^2 U_T^2} V_{LO} \frac{1}{g_{ds1} + \frac{1}{R_{IF}}} \tag{3.10}$$

The total conversion gain of the mixer including the matching network gain is expressed as follows:

$$A_{mix} = \frac{W}{L} I_{do} \frac{A_{match}^2}{n^2 U_T^2} V_{LO} \frac{1}{g_{ds1} + \frac{1}{R_{IF}}} \tag{3.11}$$

The RF channel noise of a MOS biased in subthreshold consists of mainly shot noise where the current power spectral density may be expressed as $2q I_d$ [23]. The IF noise voltage at IF1 and IF2 due to translation of the RF channel noise of M1 can be expressed as follows:

$$V_{nIF1} = \frac{\sqrt{2q I_d}}{g_m} A_{mix1} \tag{3.12}$$

$$V_{nIF2} = -NCR \cdot \frac{\sqrt{2q I_d}}{g_m} A_{mix1} \tag{3.13}$$

where NCR is noise cancellation ratio as expressed in (3.8). Since noise at IF1 and IF2 due to M1 is negatively *correlated*, the total IF noise due to M1 can be evaluated by adding (3.12) and (3.13) as *voltages* as follows:

$$V_{nIF\_M1} = (1 - NCR) \cdot \frac{\sqrt{2q I_d}}{g_m} A_{mix1} \tag{3.14}$$

Since the noise of M1 and M2 is uncorrelated the total noise voltage power spectral density at IF due to translation from RF is expressed as follows:

$$V_{n\text{IF}}^2 = V_{n\text{IF\_M1}}^2 + V_{n\text{IF\_M2}}^2 \tag{3.15}$$

As shown in (3.14), the noise voltage of each MOS is reduced after addition at IF and is $(1 - \text{NCR})$ times its original noise voltage. As a result, as shown in (3.15), the total noise power spectral density (PSD) of M1 and M2 is reduced after addition at IF and is $(1 - \text{NCR})^2$ times the original noise power spectral density. As discussed in Sect. 3.4.3, design values for NCR vary between 0.33 and 0.5 leading to 4X reduction in the effective noise power at IF.

3.5   Noise Cancellation Simulations

In order to verify the frequency translated mutual noise cancellation, SpectreRF® PSS and PNOISE simulations were used to plot the noise transfer function (NTF) from the channel noise of transistor M1 at RF to nodes V_{OUT1}, V_{OUT2}, and V_{OUT} at IF viz. NTF1, NTF2, and NTF, respectively, as shown in Fig. 3.9a, b. The NTF is the difference between NTF1 and NTF2 verifying noise cancellation of M1. Similarly, signal transfer function (STF) from PORT1 at RF to nodes V_{OUT1}, V_{OUT2}, and V_{OUT} at IF viz. STF1, STF2, and STF, respectively, has been simulated using PSS and PAC as shown in Fig. 3.9c, d. The STF is the sum of STF1 and STF2 verifying signal addition. This *noise cancellation* and *signal addition* leads to an improved noise figure. The noise figure improvement is 3 dB in simulations for the circuit in Fig. 3.9. In this prototype the simulated NCR is approximately 33% as shown in Fig. 3.9b. This is in close agreement with our analytical expression in (3.8) which shows that for a good match at input NCR = 33.33%.

3.6   Impact of Process Variation

Monte Carlo simulations were performed to quantify the effect of process variation and device mismatch on the noise figure and the gain of the receiver frontend. As shown in Fig. 3.10, the standard deviation of the noise figure at $f_{\text{LO}} = 2.4\,\text{GHz}$ is 0.086 dB due to process variation and 0.00076 dB due to device mismatch. As shown in Fig. 3.11, the standard deviation of the conversion gain at $f_{\text{LO}} = 2.4\,\text{GHz}$ is 0.073 dB due to process variation and is 0.005 dB due to device mismatch. Figure 3.12a shows simulated gain and NF vs temperature for TT, SS, and FF corners and Fig. 3.12b shows the measured variation of gain and NF over 5 samples. We note that the architecture performance is robust to process variation and device mismatch.

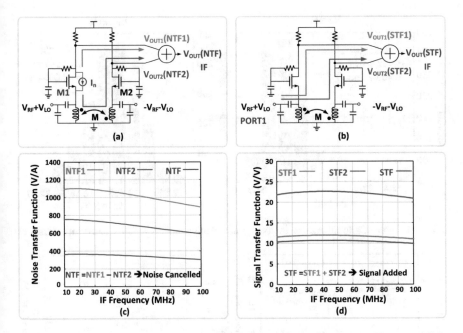

Fig. 3.9 (a) NTF paths from channel noise source of M1, (b) STF paths from PORT1, (c) NTF curves from channel noise source of M1, and (d) STF curves from PORT1

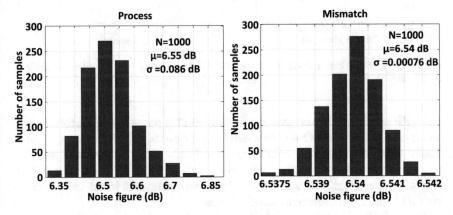

Fig. 3.10 Monte Carlo simulation results of noise figure including process variation and device mismatch

3.7 Measurement Results

The prototype was implemented in TSMC's 65 nm CMOS and occupies an area of 0.45 mm^2. The die-micrograph is shown in Fig. 3.13 and the test setup is shown in Fig. 3.14. The chip was tested by probing the die using RF probes. The RF + LO

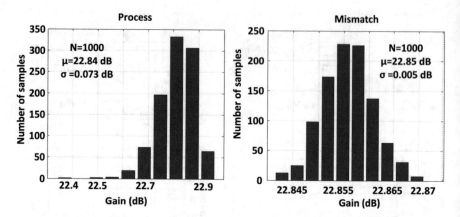

Fig. 3.11 Monte Carlo simulation results of gain including process variation and device mismatch

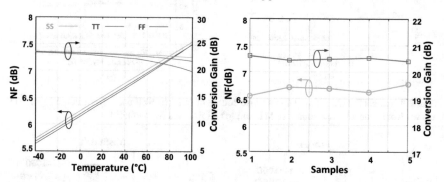

Fig. 3.12 (a) Process corner simulation of gain and NF vs temperature. (b) Measured NF and gain over 5 samples

and −RF − LO signals were generated using off-chip power combiner followed by an off-chip balun. The LO was generated using Agilent's 8257D signal generator. The S11 measurements were done using R&S ZVA67 vector network analyzer. The noise figure measurements were done using Agilent 346C noise source and R&S FSW43 spectrum analyzer with a noise figure personality. All losses have been de-embedded in the measurements provided here.

Figure 3.15a shows the measured and simulated S11 and conversion gain at a 20 MHz IF versus the RF frequency spanning 2.3–2.5 GHz. Figure 3.15b shows the measured noise figure and simulated noise figures with and without noise cancellation at a 20 MHz IF versus the RF frequency spanning 2.3–2.5 GHz. The measured noise figure curve tallies quite well with the simulated noise figure curve. The simulated noise figure is spot noise figure at 20 MHz IF. The minimum noise figure of 6.55 dB occurs at 2.5 GHz RF with a conversion gain of 20.6 dB. The noise figure at lower IF is deteriorated by the flicker noise of the buffer and adder and also

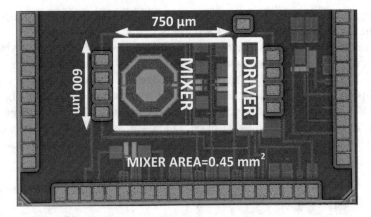

Fig. 3.13 Die-micrograph of the receiver frontend

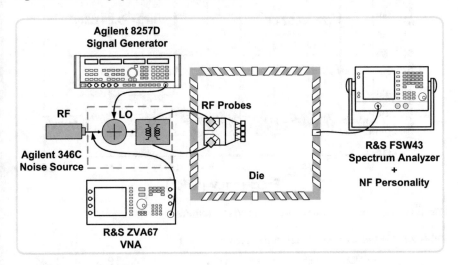

Fig. 3.14 Test setup of the receiver frontend

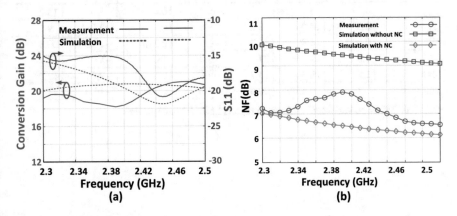

Fig. 3.15 Measured and simulated (**a**) conversion gain, S11, and (**b**) NF vs. RF frequency

due to gain reduction by the IF coupling capacitance which forms a high pass filter. The simulated spot noise figure at 1 MHz IF is 15 dB.

Figure 3.16 shows a two tone output with an IIP3 of −9.24 dBm meeting blocker specifications [2]. The performance of the prototype is summarized and compared in Table 3.1. The proposed circuit has the lowest NF while consuming the lowest power of 194 μW from a 0.7 V VDD. The proposed design has the highest FOM of 31 dB which is 10 dB higher than the state of the art [9]. Figure 3.17 shows a 3D bar chart for the various designs displaying FOM, LO power, and noise figure. It is clearly visible that the proposed design improves all three performance specifications.

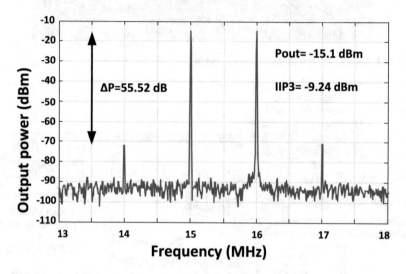

Fig. 3.16 Measured IIP3 and two tone test output spectrum

Table 3.1 Performance comparison of the receiver frontend

Reference	VDD (V)	NF[a] (dB)	Gain (dB)	IIP3 (dBm)	P_{MIX}[b] (mW)	P_{LO} (dBm)	FOM (dB)
[5]	0.77	15	5.7	−5.7	0.48	5.0	13.2
[7]	0.6	14.0	3.2	−8.0	0.8	−2.0	9.6
[9]	0.6	11.8	12.7	−6.0	0.380	−14.0	20.8
[10]	1.0	19	27	−3.0	1.0	−5	18
[24]	1.0	18.3	15.7	−9.0	0.5	−9.0	13.1
This work	0.7	6.55	20.6	−9.2	0.194	−14.0	31

[a]DSB NF [b]Mixer power except [10] which reports LNA + Mixer power

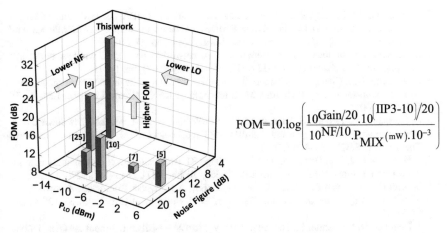

Fig. 3.17 Chart comparing FOM, LO power, and noise figure

$$FOM=10.\log\left(\frac{10^{Gain/20}.10^{\left(IIP3-10\right)/20}}{10^{NF/10}.P_{MIX}(mW).10^{-3}}\right)$$

3.8 Conclusion

We have proposed a frequency translated mutual noise cancellation technique that improves the noise figure and power consumption of transconductance mixers suitable for sub 1 V operation by exploiting the cyclostationary property of noise. Traditional noise cancellation techniques employ active devices in the auxiliary path whose noise remains uncancelled. In contrast, the proposed noise cancellation technique is mutual where one path cancels the noise of other and vice versa. This is achieved by employing a center-tapped symmetric differential inductor as a passive noise coupler which enables coupling from both ends by virtue of its reciprocity. The noise coupler behaves as a differential inductor for differential input signal, thereby contributing in input matching network but as a transformer for single ended noise, thereby contributing in noise cancellation. Furthermore, it consumes zero power and contributes negligible noise. The mixer is operational at 0.7 V with LO power of only −14 dBm. The mixer has the lowest noise figure and power consumption and its FOM is 10 dB higher than the state of the art.

References

1. M. Rahman, M. Elbadry, R. Harjani, An IEEE 802.15.6 standard compliant 2.5 nJ/Bit multiband WBAN transmitter using phase multiplexing and injection locking. IEEE J. Solid State Circuits **50**(5), 1126–1136 (2015)
2. Y.H. Liu et al., A 1.9 nJ/b 2.4 GHz multistandard (bluetooth low energy/Zigbee/IEEE802.15.6) transceiver for personal/body-area networks, in *IEEE International Solid State Circuits Conference* (IEEE, Piscataway, 2013), pp. 446–447

3. M. Rahman, M. Elbadry, R. Harjani, A 2.5 nJ/bit multiband (MBAN & ISM) transmitter for IEEE 802.15.6 based on a hybrid polyphase-MUX/ILO based modulator, in *IEEE International Solid State Circuits Conference* (IEEE, Piscataway, 2014), pp. 17–20
4. IEEE standard for local and metropolitan area networks—part 15.6: wireless body area networks, in *IEEE Std 802.15.6-2012* (IEEE, Piscataway, 2012), pp. 1–271
5. K.H. Liang, H.Y. Chang, Y.J. Chan, A 0.5–7.5 GHz ultra low-voltage low-power mixer using bulk-injection method by 0.18-um CMOS technology. IEEE Microw. Compon. Lett. **17**(7), 531–533 (2007)
6. C. Hermann, M. Tiebout, H. Klar, A 0.6-V 1.6-mW transformer-based 2.5-GHz downconversion mixer with +5.4-dB gain and -2.8-dBm IIP3 in 0.13-um CMOS. IEEE Trans. Microwave Theory Tech. **53**(2), 488–495 (2005)
7. H.H. Hsieh, L. H. Lu, Design of ultra-low-voltage RF frontends with complementary current-reused architectures. IEEE Trans. Microw. Theory Techn. **55**(7), 1445–1458 (2007)
8. V. Vidojkovic, J. van der Tang, A. Leeuwenburgh, A.H.M. van Roermund, A low-voltage folded-switching mixer in 0.18-um CMOS. IEEE J. Solid State Circuits **40**(6), 1259–1264 (2005)
9. J. Deguchi, D. Miyashita, M. Hamada, A 0.6 V 380 uW −14 dBm LO-input 2.4 GHz double-balanced current-reusing single-gate CMOS mixer with cyclic passive combiner, in *IEEE International Solid-State Circuits Conference* (IEEE, Piscataway, 2009), pp. 224–225
10. M.A. Abdelghany, R.K. Pokharel, H. Kanaya, K. Yoshida, Low-voltage low-power combined LNA-single gate mixer for 5 GHz wireless systems, in *IEEE Radio Frequency Integrated Circuits Symposium* (IEEE, Piscataway, 2011), pp. 1–4
11. H. Darabi, A.A. Abidi, Noise in RF-CMOS mixers: a simple physical model. IEEE J. Solid State Circuits **35**(1), 15–25 (2000)
12. W. Cheng, A.J. Annema, J.A. Croon, B. Nauta, Noise and nonlinearity modeling of active mixers for fast and accurate estimation. IEEE Trans. Circuits Syst. I **58**(2), 276–289 (2011)
13. S.C. Blaakmeer, E.A.M. Klumperink, D.M.W. Leenaerts, B. Nauta, Wideband balun-LNA with simultaneous output balancing, noise-canceling and distortion-canceling. IEEE J. Solid State Circuits **43**(6), 1341–1350 (2008)
14. D. Murphy et al., A blocker-tolerant, noise-cancelling receiver suitable for wideband wireless applications. IEEE J. Solid State Circuits **47**(12), 2943–2963 (2012)
15. A. Wong, M. Dawkins, G. Devita, N. Kasparidis, A. Katsiamis, O. King, F. Lauria, J. Schiff, A.J. Burdett, 1 V 5 mA multimode IEEE 802.15.6/Bluetooth low-energy WBAN transceiver for biotelemetry applications. IEEE J. Solid State Circuits **48**(1), 186–198 (2013)
16. Bluetooth SIG, *Specification of the Bluetooth System v4.0* (2010)
17. B. Razavi, *RF Microelectronics*, 2nd edn. (Prentice Hall, New York, 2011)
18. M.T. Terrovitis, R.G. Meyer, Noise in current-commutating CMOS mixers. IEEE J. Solid State Circuits **34**(6), 772–783 (1999)
19. C. Hull, *Analysis and Optimization of Monolithic RF Downconversion Receivers*. Ph.D. thesis, University of California, Berkeley, 1992
20. J.R. Long, Monolithic transformers for silicon RF IC design. IEEE J. Solid State Circuits **35**(9), 1368–1382 (2000)
21. M. Danesh, J.R. Long, Differentially driven symmetric microstrip inductors. IEEE Trans. Microw. Theory Techn. **50**(1), 332–341 (2002)
22. F. Bruccoleri, E.A.M. Klumperink, B. Nauta, Wide-band CMOS low-noise amplifier exploiting thermal noise canceling. IEEE J. Solid State Circuits **39**(2), 275–282 (2004)
23. C.C. Enz, F. Krummenacher, E.A. Vittoz, An analytical MOS transistor model valid in all regions of operation and dedicated to low-voltage and low-current applications. Analog Integr. Circ. Sig. Process **8**(1), 83–114 (1995)
24. H. Lee, S. Mohammadi, A 500 uW 2.4 GHz CMOS subthreshold mixer for ultra low power applications, in *IEEE Radio Frequency Integrated Circuits Symposium* (IEEE, Piscataway, 2007), pp. 325–328

Chapter 4
Dual-Path Noise Cancelling LNA

4.1 Introduction

In this chapter we present a 0.7 V low power LNA which combines a 1:3 frontend balun with dual-path noise and nonlinearity cancellation for improved noise performance at low powers. In traditional noise cancellation techniques only the noise of the main path is cancelled while the noise of the auxiliary path is reduced by using higher power. In the proposed design, the noise and nonlinearity of both the main and the auxiliary paths are mutually cancelled allowing for low power operation. The 2.8 dB NF, −10.7 dBm IIP3 LNA in TSMC's 65 nm GP process consumes 475 μW of power resulting in an FOM of 28.8 dB which is 8.2 dB better than the state of the art.

There is an increasing demand for low power radios with the proliferation of the internet of things (IoT), wireless body area networks (WBAN), and next generation 5G [1–3]. A WBAN is a network of medical sensors or actuators on, in, or around the human body employing wireless connectivity. WBAN is capable of revolutionizing future health care. A typical WBAN scenario includes a network of medical sensors for monitoring vital statistics such as temperature, blood sugar, and heart rate, and actuators such as insulin pumps and cardiac pacemakers. By connecting these devices with a local base unit, e.g., a cell phone, WBAN will enable remote patient monitoring by providing doctors with real-time vital data. This will result in reduced health care costs and early detection and prevention of diseases. Remote patient monitoring will significantly benefit the aging population in regions where there is a scarcity of clinics and clinicians. Furthermore, the wireless connectivity can facilitate untethered patient monitoring without limiting patient movement. However, all these devices require radios which can transmit and receive signals in order to maintain wireless connectivity. As most of these devices

© Springer Nature Switzerland AG 2020
M. Rahman, R. Harjani, *Design of Low Power Integrated Radios for Emerging Standards*, Analog Circuits and Signal Processing,
https://doi.org/10.1007/978-3-030-21333-6_4

are likely to be powered by small batteries, their radios have to be extremely power efficient. The IEEE 802.15.6 standard provides the necessary specifications for such low power radios [4].

With the proliferation of wireless technology in the last two decades, a plethora of "things" have internet connectivity. This has led to the so-called internet of things (IoT), which is a network of objects, animals, people provided with an IP address, and the ability to transfer data without human-to-human or human-to-computer interaction. Cisco has predicted that 50 billion items will be connected to the internet by 2020 [5]. As shown in Fig. 4.1 the number of connected devices per person is predicted to be 6.58 by 2020. Just as side note, the latest version of IP address mechanism, i.e., IPV6 has a sufficiently large address space to accommodate the potential exponential growth of the internet of things. However, in order to minimize the impact of such devices on the environment and on energy consumption, the power consumption of these IoT radios should be extremely low when deployed for internet connectivity. In addition, many of these nodes will not have continuous access to power and are likely to use small batteries further increasing the need for low power radios.

The rapid scaling of the power supply (Vdd) with technology reduces power consumption in digital circuits where the power consumed can be expressed as $\approx CVdd^2 f$. This facilitates the power reduction in modern radios that have large digital blocks. Therefore, it is preferable to employ sub 1V circuits in order to reduce power and accommodate technology scaling. However, a low Vdd degrades the

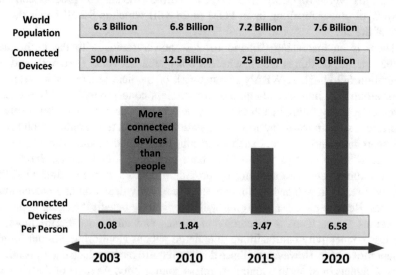

Fig. 4.1 Cisco's prediction of connected devices per person by 2020

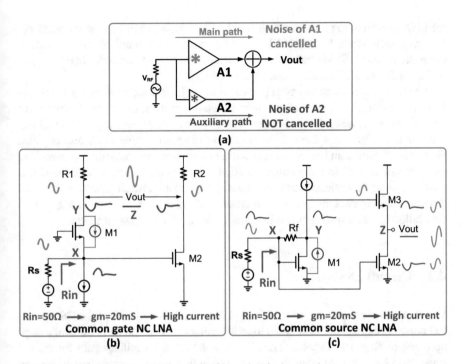

Fig. 4.2 Traditional noise cancelling (NC) LNA and their shortcomings

SNR of the RF frontend which can be partially restored by using noise cancelling techniques. Unfortunately, traditional noise cancelling LNAs are power hungry and are not well suited for low power operation [6–9].

Traditional noise cancelling techniques are based on the availability of two suitable nodes X and Y in the circuit where the signals are in phase but the noise of the main path is out of phase at these nodes as shown in Fig. 4.2a. Noise cancelling LNAs have been realized as either common source (CS) [7] or as common gate (CG) [6] amplifiers. For both these designs the input impedance is provided by an input transistor M1 as $Zin = 1/gm_1$ (i.e., common gate or high gain with resistive feedback). For a $Zin = 50\,\Omega$ the required gm is $20\,mS$ which necessitates that the power is greater than 1.5 milliamperes even for a low $\Delta V_{GS} = 150\,mV$. A larger ΔV_{GS} is required for higher linearity but results in increased power. Furthermore, the noise of the auxiliary path (M2) does not get cancelled and therefore to improve the overall NF, gm_2 is increased at the cost of higher current (Fig. 4.2b and c). Mutual noise cancellation has been reported in nonlinearity mixer-based RF frontends [10, 11] where the noise of both the main and auxiliary paths is cancelled by virtue of the mutual inductance of a differential inductor but the technique cannot be easily ported to other designs. This paper focuses on mutual noise cancellation in the LNA itself making it more versatile. The LNA design in this paper encompasses the frequency bands for the IEEE 802.15.6 narrowband standard, the industrial-scientific-medical (ISM) band, the US medical body area network (MBAN), and the European MBAN bands spanning 2.3–2.5 GHz.

A subset of this work has been presented at [12]. Here we have expanded on the work [10] with additional analysis, new simulation results, including the impact of process variation and mismatch using Monte Carlo simulations and additional test setup details. Section 4.2 provides the circuit diagram for the LNA and explains the common source and common gate noise cancellation mechanisms and matching requirement as well as simulations. Section 4.3 provides an analysis of the signal addition, noise cancellation, and nonlinearity cancellation mechanisms. Section 4.4 describes the impact of process variation and mismatch. Section 4.5 provides simulation and measurement results. Finally, Sect. 4.6 concludes the paper.

4.2 Circuit Design

We propose a dual-path noise cancelling LNA which is a hybrid of the CS and the CG noise cancelling LNAs coupled together using the two secondary turns of a step-up balun as shown in Fig. 4.3. The CG stage acts as the auxiliary path for the CS stage and vice versa. As a result, the noise of both stages gets cancelled. The input stage consists of a 1:3 balun. The secondary S1 of the balun is connected to a class AB inverter stage consisting of transistors M1a and M1b acting as a CS amplifier (A1). The secondary S2 of the balun is connected to a CG stage formed by M2 (A2). Here A1 (CS amplifier) and A2 (CG amplifier) form the two parallel amplifiers that provide gain.

4.2.1 CS Noise Cancellation

As shown in Fig. 4.4a, for noise cancellation of the CS amplifier M1a/M1b forms the main path and M2 forms the auxiliary path. The noise current of M1a and M1b flows through Rf and Rs1, where Rs1 is the impedance transformed input source resistance at S1. This results in two in-phase instantaneous noise voltages at Y1 and X1. The noise voltage at X1 is inverted at Z by M2. Note, M2 is a gm-boosted stage as the two secondary terminals of the balun, i.e., S1 and S2 provide opposite signal phases. The noise voltage at Y1 appears in phase at Z due to the source follower formed by M3. Since the noise voltages from the main and auxiliary paths are out of phase, they get cancelled at the output Z (Fig. 4.4a).

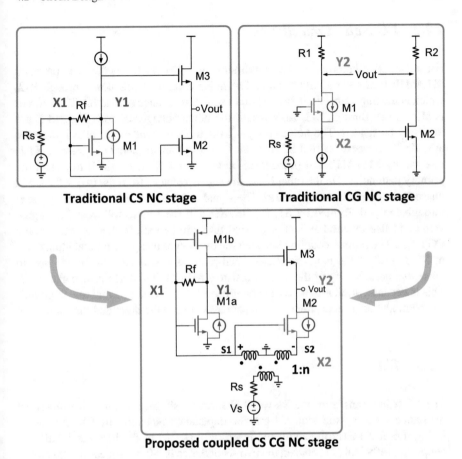

Fig. 4.3 Coupling of traditional CS and CG noise cancelling (NC) LNA stages to form a coupled CS-CG NC LNA

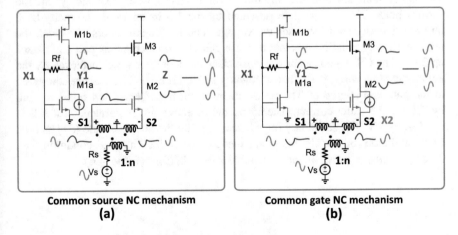

Fig. 4.4 Noise cancellation mechanism in the proposed LNA

4.2.2   CG Noise Cancellation

For noise cancellation of the CG amplifier (Fig. 4.4b) M2 forms the main path and
M1a/M1b forms the auxiliary path. The noise current of M2 flows through Rs2,
which is the impedance transformed input source resistance seen at S2. This current
is also drawn through M3. As a result, the noise voltages at node X2 and Z are
inversely correlated. The noise voltage at the input X2 undergoes phase inversion
at node X1 because S1 and S2 form an inverting transformer. This noise at X1 is
inverted by M1a/M1b and propagated by the source follower (M3) to cancel the
primary path noise. On the other hand, the signal voltages from the two paths are in
phase at the output Z and get added. The signal voltage is inverted at Z after passing
through the path formed by S1, the inverter and the source follower. The signal
voltage is also inverted at Z after passing through the path formed by S2, and the
CG stage. The mutual coupling between the two stages using the mutual inductance
of S1 and S2 of the balun facilitates dual-path noise cancellation. In addition to
noise, the nonlinearity of the input transistors M1a/M1b and M2 is also cancelled.
This is because, similar to noise, the nonlinearity in the drain current due to gm/gds
can be modelled as dependent current sources between the drain and the source [6].

4.2.3   Zin

The 1:3 balun transforms the Rs = 50 Ω source resistance to a 450 Ω differential
resistance reducing the required gm for impedance matching by 9X. As shown
in Fig. 4.6 at 2.4 GHz the simulated insertion loss (IL) of the balun is 1.4 dB and
coupling coefficient (k) between the two secondaries is −0.83. An equivalent circuit
model for the balun [13, 14] is shown in Fig. 4.5. Here, port 1 is the primary, port
2 is the inverting secondary, and port 3 is the non-inverting secondary. Mxy are
the mutual inductances and Lx are the self-inductances, where "x" and "y" are the
various ports. The voltage gain provided by the 1:3 balun would normally have
hurt the linearity as it increases the voltage seen by the transistors. However, the
nonlinearity cancellation inherent to these cancellation techniques partially makes
up for it. The CS stage is a current reuse class AB amplifier which achieves 2X the
gm on the same current. The two secondary terminals S1 and S2 that are connected
to the gate and source of M2 act as an inverting transformer and boost the gm
by 2X for the same current. Therefore, the effective gm requirement for M2 is
reduced by 9X. The secondary impedances seen by the left-hand side (S1) and the
right-hand side (S2) of the balun are made equal in the balanced condition, i.e.,
$2 \cdot gm_2 = gm_1 = gm_{1a} + gm_{1b}$ such that $Zin = 2/(9gm_1)$ (Fig. 4.6).

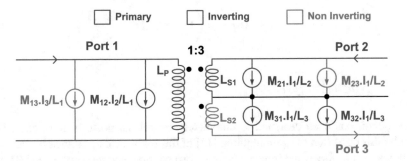

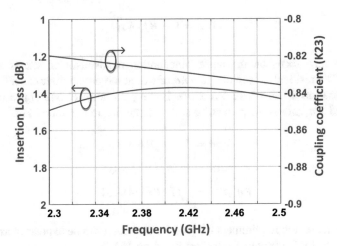

Fig. 4.5 Circuit model for the balun

Fig. 4.6 Insertion loss and coupling between secondaries of the balun

4.3 Signal, Noise, and Nonlinearity Analysis

Next we provide some theoretical analysis for signal addition and noise cancellation. This analysis is performed at RF under the assumption that the inductances associated with the balun terminals have been resonated out with appropriate capacitances, i.e., we only consider the real parts of circuits.

4.3.1 Signal Analysis

The signals through both feed-forward paths are in phase and get added to each other and can be expressed as follows, where n is the passive voltage gain from the balun.

$$Av = Av1 + Av2 = n\left(1 - gm_1 Rf + \frac{2gm_2}{gm_3}\right) \qquad (4.1)$$

4.3.2 Noise Analysis

The circuit has been designed to completely cancel the noise of M1a and M1b. Under this condition the gain required [15] of the auxiliary path formed by M2 is as follows, where Rs_1 is the impedance transformed input source resistance at S1 and $gm_1 = gm_{1a} + gm_{1b}$.

$$Av2 = 1 + Rf/Rs_1 \qquad (4.2)$$

Let us now focus on the noise cancellation mechanism of M2 as shown in Fig. 4.7. The residual noise voltage Vno of M2 at the output after cancellation can be expressed as Vno, where Vn_main is the path noise of M2 appearing at the output and Vn_aux is the path noise at the output due to M1a and M1b.

$$Vno = Vn\_main + Vn\_aux \qquad (4.3)$$

$$Vn\_main = (3In/4)(1/gm_3) \qquad (4.4)$$

$$Vn\_aux = -(InRs1/4)Av1 \qquad (4.5)$$

Furthermore, the requirement on gm_2 for matching can be expressed as follows, where the factor 2 accounts for the gm-boosting of M2.

Fig. 4.7 Simplified model
for noise cancellation of M2

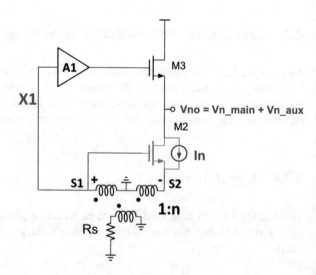

$$gm_2 = (1/2)Rs2 \tag{4.6}$$

Using these equations, the residual noise (Vno^2) of M2 at the output can be expressed as $(1/4)Vn^2$. Therefore, if the circuit is designed for 100% cancellation of A1 (M1a and M1b), then under perfect matched conditions only 75% noise cancellation of A2 (M2) can be achieved.

4.3.3   Nonlinearity Analysis

The nonlinearity of both the CS and CG transistors due to nonlinear input conductance and output conductance can be modelled as a dependent current source between the drain and the source as shown in Fig. 4.8 [6]. These current sources are a function of both the v_{gs} and v_{ds} and therefore include gm nonlinearity, gds nonlinearity, and secondary effects such as drain induced barrier lowering (DIBL). Mathematically, we can express the drain current as a function of v_{gs} using the Taylor expansion shown in Eq. (4.2). The nonlinearity terms can be summed together as shown in Eq. (4.8)

$$Ids(v_{gs}) = gm_1 v_{gs} + gm_2 v_{gs}^2 + gm_3 v_{gs}^3 + \cdots . \tag{4.7}$$

$$Ids(v_{gs}) = gm_1 v_{gs} + I_{nonlin} \tag{4.8}$$

We can similarly model nonlinearity in terms of v_{ds}. The nonlinear component of drain current undergoes the same phase reversal mechanism as noise and gets

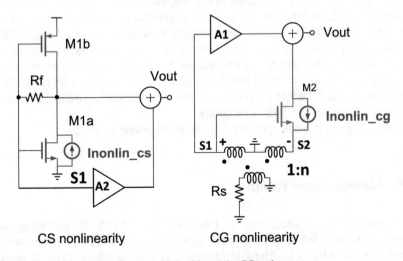

CS nonlinearity   CG nonlinearity

Fig. 4.8 Modelling nonlinearity for both the CS and the CG paths

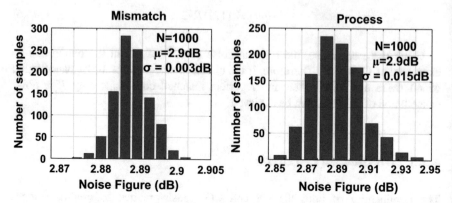

Fig. 4.9 Monte Carlo simulations for noise figure: process variation impact

cancelled in our proposed circuit. Since nonlinearity can be modelled as a current source similar to noise, we can achieve complete cancellation of nonlinearity of the CS device. Similarly, we can achieve 75% cancellation of the nonlinear terms for the CG device following the methodology used for noise cancellation derived in the previous section.

4.4   Impact of Process Variation

Monte Carlo simulations were performed using available models from the foundry to quantify the effect of process variation and device mismatch on the noise figure and the gain of the LNA. As shown in Fig. 4.9, the standard deviation of the noise figure at 2.4 GHz is 0.015 dB due to process variation and 0.003 dB due to device mismatch. As shown in Fig. 4.10, the standard deviation of the gain at 2.4 GHz is 0.1 dB due to process variation and is 0.04 dB due to device mismatch. Figure 4.11 shows simulated gain and NF vs temperature for TT, SS, and FF corners. Across temperature range of −40 °C to 100 °C, the gain variation is 2 dB at SS corner, 3 dB at TT corner, and 5.1 dB at FF corner; and the NF variation is 0.6 dB at SS corner, 0.8 dB at TT corner, and 1 dB at FF corner.

Next, we provide measurement results from the prototype design.

4.5   Measurement Results

The prototype was implemented in TSMC's 65 nm CMOS GP technology and occupies an area of 0.42 mm^2. The die-micrograph is shown in Fig. 4.12 and the test setup is shown in Fig. 4.13. The chip was tested by probing the die using RF probes. The noise figure measurements were done using an Agilent 346A noise source

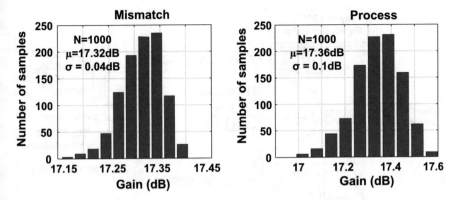

Fig. 4.10   Monte Carlo simulation results for gain: process variation impact

Fig. 4.11   Process corner simulation results for gain and noise figure vs temperature

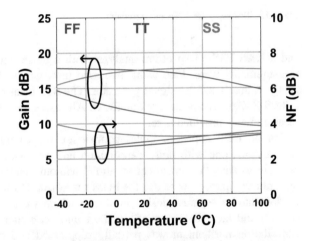

Fig. 4.12   Die-micrograph of the LNA

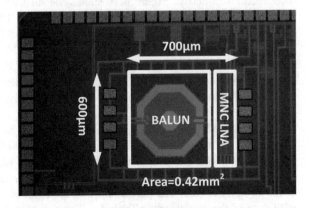

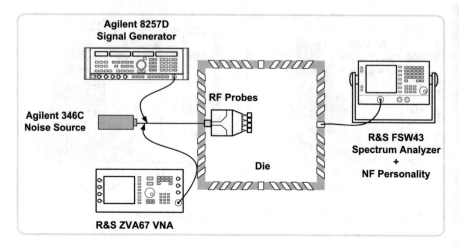

Fig. 4.13 Test setup for LNA measurement

and a R&S FSW43 spectrum analyzer with a noise figure personality. Linearity measurements were done using two tone tests where the tones were generated using Agilent's 8257D signal generator. The S11 measurements were done using a R&S ZVA67 vector network analyzer. All losses have been de-embedded in the measurements provided here.

For an ideal apples-to-apples comparison we would have preferred to have two designs, one with noise cancellation and one without. However, it was not possible to have "no noise cancellation" without changing the topology. So, the only fair comparison is to use the FOM for LNAs. However, as a compromise we implemented a separate design in silicon where we grounded the source of M2 (in A2) and kept the secondary of the balun open such that the dual-path noise cancellation mechanism was partially suppressed and only the CS cancellation mechanism was active. Figure 4.14 shows the measured and simulated NF with noise cancellation and with partial noise cancellation. The measured results match quite well with simulations in both cases. The full cancellation technique improves on the partial cancellation design by reducing the NF by 2 dB. The measured NF at 2.3 GHz is 2.8 dB. The residual NF is mainly due to insertion loss of the balun, residual noise of M2 (25%), and the uncancelled noise of Rf and M3. The noise contribution of balun, Rf, and M3 are, respectively, 29%, 6%, and 38% in simulation.

The measured gain at 2.3 GHz is 17.4 dB and the S11 is better than -13.5 dB throughout the frequency range as shown in Fig. 4.15. Figure 4.16 shows the measured two tone test output with cancellation and with partial cancellation. In comparison to partial cancellation, the full cancellation technique improves the IIP3 of the LNA by 3.4 dB. The performance of the LNA is summarized and compared with state-of-the-art designs in Table 4.1. The proposed design has the lowest power of 475 μW and the highest FOM [16] of 28.8 dB which is 8.2 dB higher than

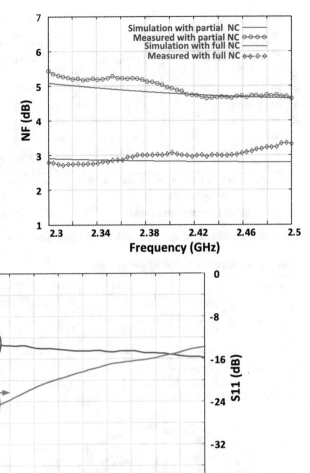

Fig. 4.14 Measured and simulated NFs with full and partial NC

Fig. 4.15 Measured gain and S11 of the LNA

the state of the art. For a fair comparison the FOM is the best parameter and an improvement in the FOM by 8.2 dB is very significant. This design has nearly 2 dB better NF and 10 dB better IIP3 than [17] while consuming half the power by virtue of the mutual noise and nonlinearity cancellation. It has similar NF, 3.7 dB lower than IIP3 but 26.5X lower power than [9]. Figure 4.17 shows a 3D bar chart of the various designs displaying the FOM, the power, and noise figures. It is clearly visible that the proposed design improves on the FOM in terms of the linearity, noise figure, and power.

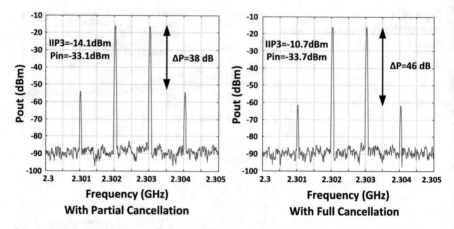

Fig. 4.16 Two tone output spectrum with full and partial cancellation

Table 4.1 Performance comparison of the LNA

Reference number	Freq. [GHz]	NF [dB]	NC [Y/N]	Gain [dB]	IIP3 [dBm]	Power [a] [mW]	FOM [dB]
[7]	2	2.4	Y	13.7	0	35	14
[6]	5.2	3.5	Y	15.6	0	14	17.8
[18]	3	2.3	N	9.8	−7	12.6	13
[17]	2.46	4.7	N	20.2	−20	0.93	20.7
Ours	2.4	2.8	Y	17.4	−10.7	0.475	28.8

[a]Includes LNA core power only

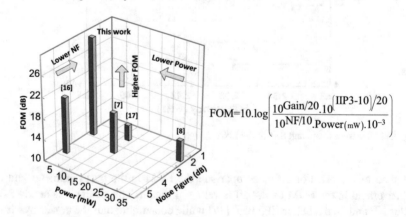

$$FOM = 10 \cdot \log\left(\frac{10^{Gain/20} \cdot 10^{\left(IIP3-10\right)/20}}{10^{NF/10} \cdot Power_{(mW)} \cdot 10^{-3}}\right)$$

Fig. 4.17 FOM, power, and noise figure comparison

4.6 Conclusions

We have proposed a dual-path mutual noise cancellation technique that improves the NF, nonlinearity, and power consumption of fully integrated LNAs with on-chip matching suitable for sub 1 V operation. In traditional noise cancellation techniques the noise of the auxiliary path is suppressed by burning significant power. In contrast, we have proposed a dual-path noise cancelling technique which mutually cancels the noise and nonlinearity of both the main (A1) and auxiliary paths (A2). This is achieved by using a balun whose secondary acts as an inverting transformer and passively couples noise from both paths by virtue of its reciprocity. The step-up balun provides voltage gain and relaxes the gm requirement for matching of the input transistors thereby reducing power. The loss in IIP3 due to passive voltage gain from the balun is compensated by nonlinearity cancellation inherent to these techniques. The circuit exploits current reuse and gm-boosting for low power. This design shows that low power noise cancellation techniques are feasible.

References

1. S. Pellerano et al., Radio architectures and circuits towards 5G, in *IEEE International Solid-State Circuits Conference* (2016), pp. 498–501
2. M. Rahman, M. Elbadry, R. Harjani, An IEEE 802.15.6 standard compliant 2.5 nJ/Bit multiband WBAN transmitter using phase multiplexing and injection locking. IEEE J. Solid State Circuits **50**(5), 1126–1136 (2015)
3. M. Rahman, R. Harjani, CMOS energy efficient integrated radios for emerging low power standards, in *IEEE Radio Frequency Integrated Circuits* (2016), pp. 151–152
4. IEEE standard for local and metropolitan area networks—part 15.6: wireless body area networks. IEEE Std. 802.15.6-2012 (2012), pp. 1–271
5. Cisco. *Cisco Internet Business Solutions Group White Paper: The Internet of Things* (2011)
6. S.C. Blaakmeer, E.A.M. Klumperink, D.M.W. Leenaerts, B. Nauta, Wideband balun-LNA with simultaneous output balancing, noise-canceling and distortion-canceling. IEEE J. Solid State Circuits **43**(6), 1341–1350 (2008)
7. F. Bruccoleri, E.A.M. Klumperink, B. Nauta, Wide-band CMOS low-noise amplifier exploiting thermal noise canceling. IEEE J. Solid State Circuits **39**(2), 275–282 (2004)
8. S.C. Blaakmeer, E.A.M. Klumperink, D.M.W. Leenaerts, B. Nauta, The blixer, a wideband balun-LNA-I/Q-mixer topology. IEEE J. Solid State Circuits **43**(12), 2706–2715 (2008)
9. D. Murphy, A. Hafez, A. Mirzaei, M. Mikhemar, H. Darabi, M.C.F. Chang, A. Abidi, A blocker-tolerant wideband noise-cancelling receiver with a 2dB noise figure, in *IEEE International Solid-State Circuits Conference* (2012), pp. 74–76
10. M. Rahman, R. Harjani, A 0.7V 194uW 31dB FOM 2.3-2.5 GHz RF frontend for WBAN with mutual noise cancellation using passive coupling, in *IEEE Radio Frequency Integrated Circuits Symposium* (2015), pp. 175–178
11. M. Rahman, R. Harjani, A Sub-1V 194μW 31-dB FOM 2.3-2.5 GHz mixer-first receiver frontend for WBAN with mutual noise cancellation. *IEEE Trans. Microwave Theory Tech.* **64**(4),1102–1109 (2016)
12. M. Rahman, R. Harjani, A sub-1v, 2.8db NF, 475μw coupled LNA for internet of things employing dual-path noise and nonlinearity cancellation, in *IEEE Radio Frequency Integrated Circuits* (2017), pp. 236–239

13. J.R. Long, Monolithic transformers for silicon RF IC design. IEEE J. Solid State Circuits **35**(9), 1368–1382 (2000)
14. J.R. Long, The modeling, characterization, and design of monolithic inductors for silicon RF IC's. IEEE J. Solid State Circuits **32**(3), 357–369 (1997)
15. F. Bruccoleri, E.A.M. Klumperink, B. Nauta, Wide-band CMOS low-noise amplifier exploiting thermal noise canceling. IEEE J. Solid State Circuits **39**(2), 275–282 (2004)
16. J. Deguchi, D. Miyashita, M. Hamada, A 0.6V 380uW-14dBm LO-input 2.4GHz double-balanced current-reusing single-gate CMOS mixer with cyclic passive combiner, in *IEEE International Solid-State Circuits Conference* (2009), pp. 224–225
17. F. Zhang, K. Wang, J. Koo, Y. Miyahara, and B. Otis, A 1.6mW 300mV-supply 2.4GHz receiver with -94dBm sensitivity for energy-harvesting applications, in *IEEE International Solid-State Circuits Conference* (IEEE, Piscataway, 2013), pp. 456–457
18. C.-W. Kim, M.-S. Kang, P.T. Anh, H.-T. Kim, S.-G. Lee, An ultra-wideband CMOS low noise amplifier for 3-5-GHz UWB system. IEEE J. Solid State Circuits **40**(2), 544–547 (2005)

Chapter 5
Transceiver

5.1 Introduction

This chapter describes an 802.15.6 compliant 2.36–2.484 GHz multiband transceiver that uses an energy efficient programmable digital power amplifier on the transit side and a zero power passive voltage gain frontend using a 1:3 balun on the receive side to achieve low power operation. A 7th harmonic injection locked oscillator and zero power passive polyphase filter generates the phases at 2.4 GHz required for phase modulation on the transmit side and for LO generation on the receive side. This enables channel selection using a 342.86 MHz PLL, i.e., at 1/7th of the RF frequency of 2.4 GHz to result in low power consumption. The prototype transmitter consumes 1.48 mW of power while delivering −9.47 dBm output power resulting in an energy efficiency of 1.52 nJ/bit at 971 kbps data rate. The measured RMS EVM for $\pi/4$ DQPSK modulation is 5.68%. The prototype receiver consumes 1.29 mW of power resulting in an energy efficiency of 1.32 nJ/bit while achieving a receiver noise figure of 10.2 dB and an IIP3 of −24.1 dBm. This design does not use off-chip inductors.

Due to the large expected number of connections in 5G and the advent of IoT and WBAN there is an increased demand for low power radios [1–3]. Unfortunately, traditional radio transmitters based on homodyne or super-heterodyne conversion schemes are power hungry due to the presence of phase locked loops (PLLs) operating at the RF frequency, need for linear mixers, and high performance data converters. Furthermore, transmitters based on polar modulation lead to more complex circuit designs. On the receive side, traditional architectures turn out to be power hungry due to the presence of the LNA at RF, linear mixers, and high performance ADCs. In this paper we present a transceiver compatible with the IEEE 802.15.6 standard [4] which employs 7th harmonic injection locking to generate phases at RF thereby drastically reducing power. The transmitter uses an energy efficient fully programmable digital power amplifier with pulse shaping capability.

© Springer Nature Switzerland AG 2020
M. Rahman, R. Harjani, *Design of Low Power Integrated Radios for Emerging Standards*, Analog Circuits and Signal Processing,
https://doi.org/10.1007/978-3-030-21333-6_5

The receiver has a zero IF architecture and uses zero power passive voltage gain and passive voltage mode mixers to substantially reduce power consumption.

5.2   System Overview

The block diagram of the complete transceiver is shown in Fig. 5.1. The transmitter consists of a digital phase-MUX-based PA which uses sinusoidal quadrature phases for PSK modulation. The sinusoidal quadrature phases are generated by a passive polyphase filter driven by an oscillator at 2.4 GHz which is injection locked to the 7th harmonic of the reference at 342.86 MHz. A pulse slimmer [5] enhances the 7th harmonic content at 2.4 GHz. Consequently, channel selection can be achieved by an integer-N PLL running at 1/7th the RF frequency which drastically reduces power consumption for frequency synthesis. In this prototype design the standard integer-N PLL needed has not been included. However, its impact on the overall architecture and power consumption has been discussed in the measurements section of the paper. The receiver is based on a zero IF I/Q architecture. It employs a passive gain stage followed by passive mixers and class AB baseband amplifiers for demodulation of the received signal.

5.3   Circuit Diagram

The circuit diagram for the transmitter is shown in Fig. 5.2. A 342.86 MHz external reference at 1/7th RF frequency is applied to a pulse slimmer. The slimmer generates pseudo differential outputs suitable for injection locking. It consists of a duty cycle control stage to enhance the 7th harmonic followed by a differentiator which suppresses lower harmonics and eliminates even order harmonics [5]. The ILO is a

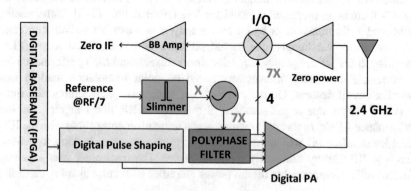

Fig. 5.1 System block diagram of the transceiver

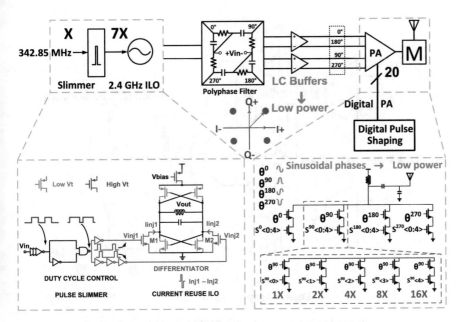

Fig. 5.2 Overall circuit details for the proposed transmitter

PMOS-NMOS current reuse oscillator which is followed by a low power LC tuned buffer. The cross-coupled pair transistors are low threshold (Vt) devices so that they contribute maximum gm for the same overdrive voltage. However, the tail current source transistor is a high threshold (Vt) device to prevent leakage due to the low threshold (Vt) devices in the cross-coupled pair. The LC buffer drives a zero power passive RC polyphase filter and generates sinusoidal quadrature I/Q phases. The quadrature phases are again buffered by a drive amplifier (DA)/LO buffer which distributes the signal to the digital power amplifier on the TX side as well as to the mixers on the RX side to serve as the LO.

The transmitter consists of a digital phase-MUX-based power amplifier which uses the sinusoidal quadrature phases to perform modulation. The transistor level circuit was shown in Fig. 5.2 but a more easily understood conceptual block diagram is shown in Fig. 5.3 for clarity. The digital amplitude control block serves both as a MUX performing phase selection and as a DAC performing amplitude control which satisfy the key requirements for phase modulation with baseband pulse shaping. Currents due to the selected phases of the appropriate amplitudes are summed at the PA output to achieve digital phase modulation. There are 4 phases and each phase has a 5 bit amplitude control resulting in 20 bit control lines. These lines are controlled by the digital pulse shaping logic which implements the square root raised cosine filter as specified in the standard. In this prototype the digital control block is implemented off-chip in an FPGA. In contrast with traditional designs [6], the PA uses sinusoidal phases generated by the polyphase filter instead of square wave phases which reduces the power consumption of the buffers driving the PA

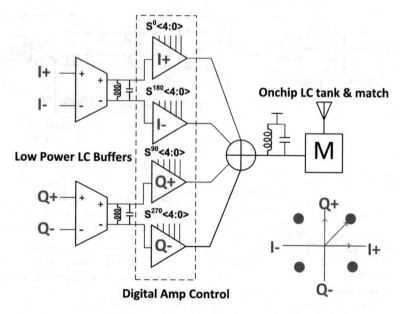

Phase MUX based PA with Digital Amplitude Control

Fig. 5.3 Conceptual circuit block diagram for the PA

and improves the spurious response. The buffers consume low power because for the same swing, the power consumption of a tuned buffer is Q/π times lower than that for CMOS inverter-based buffers where Q is the quality factor of the inductor. These buffers have a supply voltage of 0.5VDD which further reduces their power consumption. The inductive load enables a swing of 1 V peak to peak. Unlike other designs [6] an integrated matching network is included on-chip.

The receiver uses a zero IF I/Q architecture as shown in Fig. 5.4. The on-chip 1:3 step up balun provides zero power passive voltage gain. We have designed a symmetric low loss balun to minimize phase imbalance. The mixers used are passive double balanced voltage mode mixers. Unlike current mode operation, voltage mode operation does not require low input impedance transimpedance amplifiers (TIA) in the baseband which consumes significant power. Though voltage mode operation has lower linearity it is sufficient for this standard. The baseband amplifiers operate in class AB mode to achieve low power and sufficient linearity.

5.4 Measurement Results

The prototype transceiver was implemented in Global Foundry's 130 nm CMOS technology and occupies an area of 1.57 mm^2. The die-micrograph is shown in Fig. 5.5. The measured RMS EVM for $\pi/4$ DQPSK modulation is 5.68% as shown

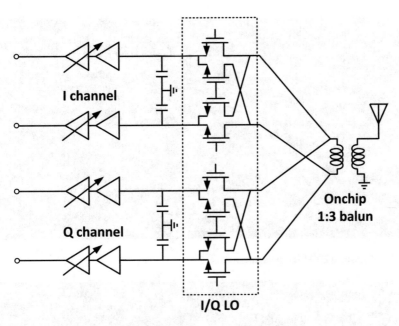

Fig. 5.4 Receiver frontend circuit details

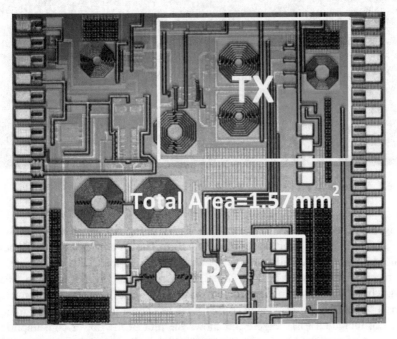

Fig. 5.5 Die-micrograph for the proposed transceiver

Fig. 5.6 Measured EVM for the transmitter

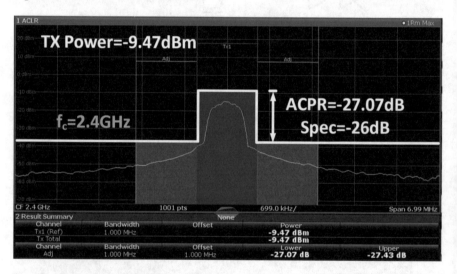

Fig. 5.7 Measured ACPR for the transmitter

in Fig. 5.6. The transmitter output mask is shown in Fig. 5.7 which shows an ACPR of -27.07 dB at a transmit power of -9.47 dBm. The PA consumes $520\,\mu$W power while delivering -9.47 dBm output power to a 50 Ω load. The noise figure of the receiver chain is 10.2 dB at 2.36 GHz as shown in Fig. 5.8 and the IIP3 is -24.1 dBm which is sufficient for the standard (-19 dBm [7]). The conversion gain is 62.7 dB at RF frequency of 2.36 GHz as shown in Fig. 5.8. The prototype does not include a PLL. Figure 5.9 shows the plot of the power consumption vs frequency of recent PLLs in the literature which indicates an increasing trend of power at higher frequencies. The plot validates the viability of a 342.86 MHz PLL

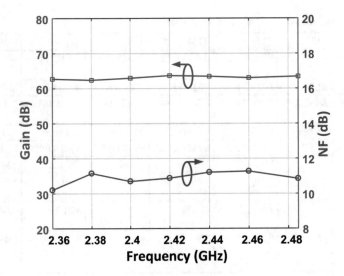

Fig. 5.8 Measured gain and noise figure for the receiver

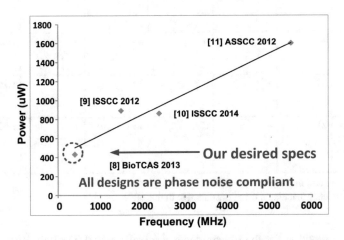

Fig. 5.9 Power vs frequency of existing PLLs in the literature

consuming approximately 450 μW. The phase noise requirement for the PLL at
2.4 GHz for this standard is −94.26 dBc/Hz at 1 MHz offset. Therefore, the phase
noise specification for the 342.86 MHz PLL will have to be 20 log(7) dB lower
due to 7th harmonic lock, i.e., −103.8 dBc/Hz. In order to achieve a reasonable
power estimate only PLLs satisfying this phase noise criterion [8–11] have been
included in Fig. 5.9. The required channel spacing and reference frequency for this
integer-N PLL is 1 MHz/7, i.e., 142.8 KHz due to 7th harmonic lock. A 10 MHz
crystal followed by divide by 70, consuming a few microwatts of power, may be
used for reference generation. After including the estimated PLL power the total

Table 5.1 Performance comparison of the transceiver

References	[12]	[13]	[14]	[15]	[1]	Ours [16]
Publication title	–	ISSCC'12	ISSCC'12	ISSCC'13	RFIC'14	–
Standard	BLE	802.15.6	802.15.6	802.15.6	802.15.6	802.15.6
Frequency (GHz)	2.4–2.484	2.36–2.4	2.36–2.484	2.3–2.484	2.3–2.484	2.3–2.484
Data rate	1 Mbps	971 kbps	971 kbps	971 kbps	971 kbps	971 kbps
Modulation	GFSK	$\pi/2$DBPSK	$\pi/4$DQPSK	$\pi/4$DQPSK	$\pi/4$DQPSK	$\pi/4$DQPSK
Technology	–	90 nm	130 nm	90 nm	130 nm	130 nm
TX Pout (dBm)	0	−10	−10	−10	−10	−9.47
TX EVM (%)	–	7.6%	10.1%	7.3%	3.21%	5.68%
TX power [a] (mW)	17.8	5.2	5.9	4.6	3	1.48
TX energy efficiency [b] (nJ/bit)	17.8	5.35	6	4.73	3.1	1.52
RX noise figure (dB)	–	–	6	6	–	10.2
RX power [a] (mW)	21	–	6.5	3.8	–	1.29
RX energy efficiency [b] (nJ/bit)	21	–	6.7	3.91	–	1.32

[a] Total power excluding digital baseband power
[b] Energy efficiency calculated using the data rate in this table

power consumption for the transmitter is 1.48 mW with 1.52 nJ/bit efficiency and that for the receiver is 1.29 mW with 1.32 nJ/bit efficiency. The performance of the prototype is summarized and compared with state-of-the-art designs in Table 5.1. The design has the lowest power and highest efficiency per bit. The noise figure is sufficient for the standard.

5.5   Conclusion

We have proposed a new architecture for a fully integrated low power transceiver compatible with the IEEE 802.15.6 standard which employs 7th harmonic injection locking. The design has no off-chip inductors. The transmitter uses an energy efficient programmable digital power amplifier. The power amplifier uses quadrature phases generated by a simpler passive polyphase filter as compared to our previous design. The simpler polyphase filter results in less signal loss and lower buffer power. The receiver uses passive voltage gain and passive voltage mixers to reduce power. This design is simpler and more robust in comparison to standard polar modulation schemes. The transceiver is more energy efficient than the state of the art by a factor of 2X on the transmit side [1] and 3X on the receive side [15]. Traditionally, high data rate transceivers with complex architectures and higher order modulation schemes result in high spectral efficiency. The new simple architecture and design changes made here allow us to approach the 1 nJ/bit limit for low power transceivers. Though the design was made to be IEEE 802.15.6 compliant, much of design and circuit architecture is easily portable to emerging low power standards.

References

1. M. Rahman, M. Elbadry, R. Harjani, A 2.5nJ/bit multiband (MBAN & ISM) transmitter for IEEE 802.15.6 based on a hybrid polyphase-MUX/ILO based modulator, in *IEEE Radio Frequency Integrated Circuits* (2014), pp. 17–20
2. S. Pellerano et al., Radio architectures and circuits towards 5G, in *IEEE International Solid-State Circuits Conference* (2016), pp. 498–501
3. M. Rahman, R. Harjani, A 0.7V 194uW 31dB FOM 2.3–2.5 GHz RF frontend for WBAN with mutual noise cancellation using passive coupling, in *IEEE Radio Frequency Integrated Circuits Symposium* (2015), pp. 175–178
4. IEEE Standard for Local and Metropolitan Area Networks - Part 15.6: Wireless Body Area Networks. *IEEE Std 802.15.6-2012* (2012), pp. 1–271
5. M. Rahman, M. Elbadry, R. Harjani, An IEEE 802.15.6 standard compliant 2.5 nJ/Bit multiband WBAN transmitter using phase multiplexing and injection locking. IEEE J. Solid State Circuits **50**(5), 1126–1136 (2015)
6. X. Liu, M.M. Izad, L. Yao, C.H. Heng, A 13 pJ/bit 900 MHz QPSK/16-QAM band shaped transmitter based on injection locking and digital PA for biomedical applications. IEEE J. Solid State Circuits **49**(11), 2408–2421 (2014)
7. M. Rahman, R. Harjani, A Sub-1V 194μW 31-dB FOM 2.3-2.5 GHz mixer-first receiver frontend for WBAN with mutual noise cancellation. IEEE Trans. Microwave Theory Tech. **64**(4), 1102–1109 (2016)
8. J. Yang, E. Skafidas, A low power MICS band phase-locked loop for high resolution retinal prosthesis. IEEE BioTCAS **7**(4), 513–525 (2013)
9. A. Elshazly et al., A 1.5GHz 890w digital MDLL with 400fsrms integrated jitter, 55.6 dbc reference spur and 20 fs/mv supply-noise sensitivity using 1b TDC, in *IEEE International Solid-State Circuits Conference* (2012), pp. 242-244
10. V. K. Chillara et al., An 860w 2.1-to-2.7GHz all-digital PLL-based frequency modulator with a DTC-assisted snapshot TDC for WPAN (Bluetooth Smart and Zigbee) applications, in *IEEE International Solid-State Circuits Conference* (2014), pp. 172–173
11. S. Ikeda et al., A 0.5-V 5.5-GHz class-C-VCO-based PLL with ultra-low-power ILFO in 65 nm CMOS, in *IEEE Asian Solid-State Circuits Conference* (2012), pp. 357–360
12. Nordic Semi, nRF8001 Preliminary Product Specification-Bluetooth Low Energy (2011)

13. Y.H. Liu et al., A 2.7nJ/b multi-standard 2.3/2.4GHz polar transmitter for wireless sensor networks, in *IEEE International Solid-State Circuits Conference* (2012), pp. 448–450
14. A. Wong, M. Dawkins, G. Devita, N. Kasparidis, A. Katsiamis, O. King, F. Lauria, J. Schiff, A. Burdett, A 1 V 5 mA multimode IEEE 802.15.6/Bluetooth low-energy WBAN transceiver for biotelemetry applications, in *IEEE International Solid State Circuits Conference* (2012), pp. 300–301
15. Y. H. Liu et al., A 1.9nJ/b 2.4GHz multistandard (bluetooth low energy/zigbee/IEEE802.15.6) transceiver for personal/body-area networks, in *IEEE International Solid-State Circuits Conference* (2013), pp. 446–447
16. M. Rahman, R. Harjani, A 2.4GHz IEEE 802.15.6 compliant 1.52nJ/bit TX amp; 1.32nJ/bit RX multiband transceiver for low power standards, in *IEEE International Conference on Electronics Circuits and Systems* (2018), pp. 821–824

Chapter 6
Conclusions

The techniques described in this book have immense potential to satisfy the quest for ever increasing power constraint in ultra low power radios for the emerging internet of things and wireless body area networks. The primary impact of these techniques will be on next-generation low power wireless consumer electronics where the data rates will multiply but the power budget for radio will be more constrained demanding longer battery life. Sub-harmonic injection locking is a powerful technique to reduce power consumption of the frequency synthesizer using PLL running at a very low frequency. A higher harmonic may be chosen to run the PLL at even lesser frequency thereby further reducing the power of the synthesizer. This also allows modulation at lower frequency further reducing transmit power consumption. The noise cancelling techniques used in the mixer and LNA are key to achieving low voltage operation of the receiver frontend with excellent figure of merit. Zero power passive voltage gain receiver frontend using a 1:3 balun combined with passive mixer-based downconversion is a simple and robust architecture which allows us to approach the 1 nJ/bit limit for low power transceivers. The digital power amplifier is highly flexible in terms of types of modulation and may be upgraded to encompass broader modulation capabilities for internet of things. Recently, radio standards for internet of things are being drafted for operation at different frequency bands. Although the design and circuit techniques described in this book were for an IEEE 802.15.6 standard compliant radio, much of them may be easily portable to emerging low power standards.

© Springer Nature Switzerland AG 2020   67
M. Rahman, R. Harjani, *Design of Low Power Integrated Radios for Emerging
Standards*, Analog Circuits and Signal Processing,
https://doi.org/10.1007/978-3-030-21333-6_6

Index

A
Agilent 346A noise source, 52
Armstrong, Edwin, 3

C
Circuits
  class-AB power amplifier (PA), 12
  eight-phase polyphase filter and MUX, 10
  fixed phase selection ILO (FPS-ILO)
      technique, 14, 15
  ILO and plot, 11–12
  layout techniques, 15
  low noise amplifier (LNA)
    CG noise cancellation, 45, 46
    CS noise cancellation, 44, 45
    zin, 46–47
  MBAN/ISM standard, 15
  PA and matching network, 14
  PA nonlinearity specification, 9–10
  phase mapping, 14–15
  phase transitions, 11
  PLL phase noise, 8–9
  PMOS-NMOS current reuse oscillator, 13
  pulse slimmer and ILO, 12
  3rd harmonic content, 13
  receiver
    FTMNC mixer with signal addition,
      27–28
    noise cancellation ratio, 29–32
    noise path, 29
    on-chip matching network, 27
    signal path, 28

  traditional injection locked technique *vs.*
      proposed technique, 14
  transceiver
    cross-coupled pair transistors, 59
    digital amplitude control block, 59
    digital phase-MUX-based power
      amplifier, 59
    342.86MHz external reference, 58
    receiver frontend circuit, 61
    transimpedance amplifiers (TIA), 60
    voltage mode operation, 60

D
Digital power amplifier, 67

E
Energy consumption, 2

F
Frequency translated mutual noise cancelling
    (FTMNC), 25

I
IEEE 802.15.6 standard, 1, 3–7, 9, 19, 23–27,
    42, 43, 57, 64, 65, 67
Injection locking, 3, 6, 21, 57, 58, 67
Internet prediction of connected things/devices,
    3
Internet of things (IoT), 1, 2, 42, 67

© Springer Nature Switzerland AG 2020
M. Rahman, R. Harjani, *Design of Low Power Integrated Radios for Emerging
Standards*, Analog Circuits and Signal Processing,
https://doi.org/10.1007/978-3-030-21333-6

L
Low noise amplifier (LNA), 4
    circuit design
        CG noise cancellation, 45, 46
        CS noise cancellation, 44, 45
        zin, 46–47
    Cisco's prediction of connected devices, 42
    IEEE 802.15.6 standard, 42
    internet of things (IoT), 42
    measurement results
        Agilent 346A noise source, 52
        Die-micrograph, 51
        FOM, 53, 54
        performance comparison, 54
        power and noise figure comparison, 54
        process corner simulation results, 51
        R&S FSW43 spectrum analyzer, 52
        S11 measurements, 52
        test setup for, 52
        two tone output spectrum, with full and
            partial cancellation, 54
    Monte Carlo simulations, 44
    noise analysis, 48–49
    nonlinearity analysis, 49–50
    process variation, impact of, 50
    remote patient monitoring, 41
    signal analysis, 47–48
Low power radios, 2
Low power receiver, *see* Receiver
Low power transmitter, *see* Transmitter

M
Monte Carlo simulations, 16–17, 34, 44, 50, 51

N
Noise analysis, 48–49
    conversion gain of mixer, 33
    RF channel noise, 33
Noise cancellation, 4, 67
    dual-path noise, LNA (*see* Low noise
        amplifier)
    M1's single ended noise current, 29, 31
    NCR simulation *vs.* upconverted source
        resistance, 32
    noise current division model, 29, 31
    simulations, receiver, 34
Nonlinearity analysis, 49–50

P
π/4 DQPSK modulation, 6, 8, 11, 26, 57, 60
Phase locked loops (PLLs), 57

Physical layer service data unit (PSDU), 27
Power consumption, 2, 4

R
Radio frequency (RF), 33, 34, 36–37, 43, 47,
        50, 57, 58
Receiver, 4
    circuit design
        FTMNC mixer with signal addition,
            27–28
        noise cancellation ratio, 29–32
        noise path, 29
        on-chip matching network, 27
        signal path, 28
    circuit level specifications
        linearity, 27
        noise figure, 26–27
    frequency translated mutual noise
        cancelling (FTMNC), 25
    IEEE 802.15.6 standard, 23–24
    low VDD operation, 24
    measurement results, 35–39
    noise analysis, 33–34
    noise cancellation simulations, 34
    passive noise coupling mechanism, 25
    process variation, impact of, 34–36
    remote patient monitoring, 23
    system level specifications, 26
    traditional noise cancellation technique, 24
    traditional switching mixer, 24
    WBAN, 23
R&S FSW43 spectrum analyzer, 52

S
Signal analysis, 47–48
Sub-harmonic injection locking, 67

T
Transceiver, 4
    circuit diagram
        cross-coupled pair transistors, 59
        digital amplitude control block, 59
        digital phase-MUX-based power
            amplifier, 59
        342.86MHz external reference, 58
        receiver frontend circuit, 61
        transimpedance amplifiers (TIA), 60
        voltage mode operation, 60
    digital phase-MUX-based PA, 58
    homodynesuper-heterodyne conversion
        schemes, 57

measurement results
    ACPR, 62
    die-micrograph, 60, 61
    EVM, 62
    noise figure of receiver chain, 62
    performance of prototype, 64
    power consumption *vs.* frequency of
        PLL, 62–63
    phase locked loops (PLLs), 57
    power amplifier, 65
    sinusoidal quadrature phases, 58
    system block diagram, 58
Transimpedance amplifiers (TIA), 60
Transmission bandwidth, 3
Transmitter, 4
    analysis, 15–16
    circuit level specifications
        PA nonlinearity specification, 9–10
        PLL phase noise, 8–9
    home automation and infrastructure
        monitoring application, 5
    IEEE 802.15.6 protocol, 5
    ILO self-resonance frequency, 6
    measurements
        EVM and ACPR measurements, 17
        ILO phase noise, 20
        800MHz reference signal, 17
        output transmit spectrum, 19

        power consumption distribution, 17–18
        power consumption *vs.* operating
           frequency, 20
        power *vs.* frequency of existing PLLs,
           20
        TX output spectrum, 19
        wideband transmit mask, 19
        wideband TX output spectrum and
           transmit mask, 19
    performance comparison, 21
    process variation and device mismatches,
        16–17
    proposed low power transmitter, 7
    RF phase multiplexing, 5
    ring oscillator-based phases, 6
    self-resonance frequency, 6
    standard PLL design, 6
    system level specifications, 7–8
    traditional mixer-based up-conversion
        transmitters, 5

W
Wireless body area network (WBAN), 1, 2, 23,
    41, 57, 67
    IEEE 801.15.6 standard, 4
    patient monitoring schemes, 1
    untethered patient monitoring, 2

Printed in the United States
By Bookmasters